Student Study Guide

to accompany

Human Biology

Ninth Edition

Sylvia S. Mader

Boston   Burr Ridge, IL   Dubuque, IA   Madison, WI   New York   San Francisco   St. Louis
Bangkok   Bogotá   Caracas   Kuala Lumpur   Lisbon   London   Madrid   Mexico City
Milan   Montreal   New Delhi   Santiago   Seoul   Singapore   Sydney   Taipei   Toronto

The **McGraw·Hill** Companies

Student Study Guide to accompany
HUMAN BIOLOGY, NINTH EDITION
SYLVIA S. MADER

Published by McGraw-Hill Higher Education, an imprint of The McGraw-Hill Companies, Inc., 1221 Avenue of the Americas,
New York, NY 10020. Copyright © 2006 by The McGraw-Hill Companies, Inc. All rights reserved.

 This book is printed on recycled, acid-free paper containing 10% postconsumer waste.

2 3 4 5 6 7 8 9 0 QPD/QPD 0 9 8 7 6
ISBN-13: 978-0-07-293615-5
ISBN-10: 0-07-293615-0

www.mhhe.com

CONTENTS

THE TEXTBOOK

Human Biology, ninth edition, by Sylvia S. Mader has many student aids that you should be sure to use.

Text Introduction: Chapter 1 discusses the characteristics of humans and presents an overview of the book. It outlines the major biological principles and reviews the scientific method that allowed scientists to arrive at these principles.

Part Introduction: An introduction to each part highlights the central ideas of that part and specifically tells you how the topics within each part contribute to biological knowledge.

Chapter Concepts: Each chapter begins with a list of concepts that organize the content of the chapter into a few meaningful questions. The concepts provide a framework for the content of the chapter.

The concepts are grouped under the major sections of the chapter and are page referenced for study. This numbering system, used in the text material and in the summaries, allows you to study the chapter in terms of the concepts presented.

Key Terms: Key terms are boldfaced in the chapter and are defined in context. These terms are page referenced in *Understanding Key Terms* at the end of the chapter and are defined in the glossary.

Internal Summary Statements: Short internal summary statements appear at the end of major sections and help you to focus your study efforts on the basics. Such statements are set off for easy identification.

Readings: Reading topics correlate to the subject matter of the chapter. Three types of readings are included in the text:

Health Focus readings review measures to keep healthy, such as the need for exercise ("Exercise, Exercise, Exercise," Chapter 11) and guidelines for preventing STDs ("Preventing Transmission of STDs," Chapter 23).

Ecology Focus readings draw attention to a particular environmental problem, such as acid rain ("The Harm Done by Acid Rain," Chapter 2)

Bioethical Focus readings present bioethical topics to stimulate thought and discussion about today's most relevant bioethical issues, such as whether performance-enhancing drugs should be used by athletes ("Performance-Enhancement," Chapter 11).

Illustrations: The illustrations in *Human Biology* are consistent with multicultural educational goals.

Integrative illustrations relate micrographs with drawings. This enables students to see an actual structure alongside a diagram drawn similarly to the micrograph. Leaders and labels common to both the micrograph and the drawing allow the student to accurately locate and visualize a particular structure.

The *Visual Focus* illustrations show in depth a main topic, system, or cycle in the chapter.

Chapter Summaries: The summary, called *Summarizing the Concepts,* is organized according to the major sections of the chapter, and the content of each section is summarized in a short paragraph or two. Chapter summaries offer a concise review of material in each chapter. You may read them before beginning the chapter to preview the topics of importance, and you may also use them to refresh your memory after you have a firm grasp of the concepts presented in each chapter.

Chapter Questions: *Studying the Concepts* are page-referenced short-answer questions that follow the organization of the material in the chapter. *Thinking Critically About the Concepts* are thought questions that encourage you to apply the chapter concepts to the real-life story that opened the chapter. *Testing Your Knowledge of the Concepts* are objective questions that allow you to test your ability to answer recall-based questions. At least one question requires that you label a diagram or fill in a table. Answers to *Testing Your Knowledge of the Concepts* appear in Appendix A of your textbook.

Online Learning Center: This section alerts you to the additional resources available on the Online Learning Center (www.mhhe.com/maderhuman9).

Other

Further Readings: If you would like more information about a particular topic or are seeking references for a research paper, the Online Learning Center has a list of current articles and books to help you get started. Entries consist of articles from *Scientific American* and other popular scientific journals and from specialty books that expand on the topics covered in the chapter.

Appendix: Appendix A, "Answer Key", provides answers to the *Testing Your Knowledge of the Concepts* questions.

Glossary: The text glossary defines the terms most necessary for making the study of biology successful. With this tool, you can review the most frequently used terms. Pronunciations are given for words of greater difficulty.

Index: The text concludes with an index so that you may quickly locate the page or pages on which any given topic is discussed.

TO THE STUDENT

This *Study Guide* was designed for the student and written to accompany your text, *Human Biology.* There is a study guide chapter for each corresponding text chapter. Each chapter contains the features listed here.

Study Tips: The study tips help you plan the best way to study the chapter. Concepts that are easily confused and terminology that is used for several purposes are also explained in this section.

Study Questions: The study questions are referenced to the sections of the text chapter and therefore can be answered as you study each section. They are also correlated to the study guide objectives. The study questions challenge you to read the appropriate section of the text and use this information to answer the questions. They provide you with the active practice you need to learn, understand, and apply the concepts in the text.

Definitions: Your study of human biology will likely require you to learn many new words. New terminology is often what frustrates students the most. To help you enjoy learning new terms, a Definitions Crossword or Definitions Wordsearch has been included at the end of the study questions in each chapter. Each chapter in your textbook has an end-of-chapter glossary of selected key terms. The definitions used for the crossword and wordsearch puzzles are taken from the "Understanding Key Terms" section. Study the terms in your textbook first; then use the crossword and wordsearch puzzles as a means of testing your knowledge.

Chapter Test: Most students recognize that there is a big difference between studying a chapter and taking a test on the chapter. This section is designed to help you overcome this hurdle. Taking this test just as you would an actual test will help you determine if you are ready to be tested by your instructor.

The chapter test is divided into two parts: the objective test and the thought questions. The objective test is a series of multiple-choice test questions (and perhaps some true/false questions) that are answered by simply selecting the correct answer. *The questions in the objective test are sequenced according to the objectives.* The thought questions require that you write out your answer in complete sentences. In many institutions, biology instructors are placing a new emphasis on developing critical thinking skills. This section will assist you in developing your writing and critical thinking skills.

Answer Key: Each chapter ends with an answer key, which contains answers for the study questions, definitions, and chapter test. Using the answer key can help you determine how well you are meeting the objectives of the chapter.

1

A HUMAN PERSPECTIVE

By entering into a study of human biology, you will learn about the structure and function of the body's systems. Therefore, you will be better equipped to make wise decisions about the health of yourself and your family members. You will also be learning about the evolution of humans and their place in the biosphere. Many human activities have impacted ecosystems for the worse, and your study should help you to make choices that will help conserve ecosystems from now on.

This chapter introduces you to these topics, and it describes the scientific method, by which scientists have discovered data that pertain to all aspects of human biology. As you study each chapter, read the integrated outline and then read the chapter, paying close attention to the boldface terms and summary statements. After each section, answer the questions at the end of the chapter for that section and answer the study guide questions for that section. Finally, take the chapter test in both the text and the study guide. Answers to Testing Your Knowledge of the Concepts in the text are found in Appendix A. Answers to all questions in the study guide are located at the end of each study guide chapter.

TIP: Be sure to visit the Online Learning Center that accompanies *Human Biology* 9/e. It has practice quizzes, interactive activities, labeling exercises, art quizzes, animations, flash cards, and much more. http://www.mhhe.com/maderhuman9

STUDY QUESTIONS

Study the text section by section. Answer the study questions so that you can fulfill the learning objectives for each section.

1.1 BIOLOGICALLY SPEAKING (PAGES 2–4)

After you have answered the questions for this section, you should be able to
- Describe the biological characteristics that define human beings.
- Recognize how humans fit into the world of living things.
- Realize that many human activities threaten the biosphere.

1. The bodies of multicellular creatures, such as humans, are highly _____.

2. Humans are made up of cells, grouped into _____, and then organs.

3. When humans _____, they create a copy of themselves and perpetuate the species.

4. We are biological organisms and share traits with all other living things, but we are different from all others because of our cultural heritage. What is meant by "cultural heritage"?

   _____

   _____

5. Humans are the product of _____, an ongoing process that has occurred over millions of years.

6. _____, the group to which humans belong, have a nerve cord protected by a vertebral column.

7. Humans belong to the animal kingdom in the domain Eukarya. List the other three kingdoms in this domain.
   a._____ b._____ c._____

8. We are a part of the _____, the layer of life that surrounds the Earth.

9. All organisms of the same species belong to a(n) _____ that inhabits a given area.

10. The total number of species in the biosphere is its _____.

11. By altering the natural ecosystems by human activities, we are causing the _____ of many species of organisms.

1.2 THE PROCESS OF SCIENCE (PAGES 6–9)

After you have answered the questions for this section, you should be able to
- Understand how the scientific process is used to gather information and to arrive at conclusions.
- List the steps of the scientific method.

12. Match the descriptions to these theories:

cell theory   theory of homeostasis   theory of evolution   gene theory   ecosystem theory

a. _____ Common descent with modification of form.

b. _____ All organisms are composed of cells.

c. _____ Organisms inherit coded information.

d. _____ Organism's internal environment stays relatively constant.

e. _____ Interaction of organisms within a particular locale.

13. Use these terms to complete the adjacent diagram:
conclusion
experiment/observations
hypothesis
observation
scientific theory

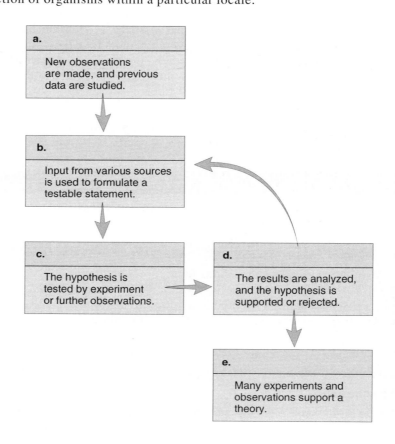

a.

New observations are made, and previous data are studied.

b.

Input from various sources is used to formulate a testable statement.

c.

The hypothesis is tested by experiment or further observations.

d.

The results are analyzed, and the hypothesis is supported or rejected.

e.

Many experiments and observations support a theory.

14. Scientists performed a controlled laboratory study. Fill in the blanks using these terms:
conclusion   the control group   data   hypothesis
a. The _____ were used to construct a graph.
b. The first _____ was not supported by the experimental results.
c. The group of mice that did not receive any substance S in their diet is called _____.
d. On the basis of further experimental results, scientists came to the _____ that sweetener S is safe if the diet contains a limited amount.

After you have answered the questions for this section, you should be able to
- Realize that all persons have the responsibility to decide how scientific information can best be used to make ethical or moral decisions.

In questions 15–16, choose the best answer.

15. Who is responsible for deciding how we use the fruits of science?
    a. scientists
    b. laypeople
    c. ministers
    d. All of the above are correct.
    e. Only *b* and *c* are correct.

16. Which of these is a value judgment?
    a. If the biodiversity of plants declines, we may miss the opportunity to find cures for various illnesses.
    b. It's wrong to kill plants because they are defenseless against humans.
    c. Due to the use of cars for transportation, air pollution has increased.
    d. People shouldn't use their cars because it increases air pollution.
    e. Only *b* and *d* are value judgments.

DEFINITIONS CROSSWORD

Review key terms by completing this crossword puzzle, using the following alphabetized list of terms:

biodiversity
biosphere
cell
control
data
evolution
homeostasis
hypothesis
kingdom
producer
reproduce
theory
vertebrates

Across
1   The total number of species within the biosphere.
3   A major scientific _____ is dependent on much experimentation.
7   Humans are in the _____ Animalia.
8   Species _____ to make offspring.
10 A scientific theory in biology that suggests that all organisms arose from earlier forms.
11 What a scientist tests in an experiment.

Down
1   The layer of life surrounding this planet.
2   What the scientist gathers when making observations.
4   Absorbs solar energy and uses it to carry on photosynthesis.
5   Organ systems help the body maintain _____, or steady state.
6   Humans and other organisms with a nerve cord protected by bone are _____.
9   The classification category following domain.
12 The basic unit of life of all living things.

OBJECTIVE TEST

Do not refer to the text when taking this test.

_____ 1. Choose the highest level of organization.
   a. cells
   b. tissues
   c. organs
   d. organ systems

_____ 2. The basic unit of all living things is the
   a. nucleus.
   b. molecule.
   c. cell.
   d. atom.

_____ 3. Organ systems of the body operate together to help maintain
   a. homeostasis.
   b. body temperature.
   c. reproductive status.
   d. circulation.

_____ 4. Which organ system functions to convert food particles to nutrient molecules?
   a. nervous
   b. digestive
   c. respiratory
   d. immune

_____ 5. Unlike other organisms on Earth, humans have
   a. a large brain.
   b. a vertebral column.
   c. manual dexterity.
   d. a cultural heritage.

_____ 6. To which of the following animals are humans most closely related?
   a. dogs
   b. chimpanzees
   c. wolves
   d. monkeys

_____ 7. All of the members of the same species that occupy an area at the same time constitute an
   a. community.
   b. ecosystem.
   c. population.
   d. tribe.

_____ 8. Plants are unique among living things in that they are
   a. multicellular and absorb food.
   b. unicellular and eat food.
   c. multicellular and produce food.
   d. All of these are correct.

_____ 9. Which of the following statements is false?
   a. More ecosystems undergo modifications as the human population increases in size.
   b. Freshwater ecosystems provide us with drinking water.
   c. The biosphere needs to be preserved so that humans can benefit.
   d. As ecosystems are destroyed, the number of species living there will increase.
   e. Science does not make moral decisions.

_____10. The process of science can best be described as
   a. objective.
   b. subjective.
   c. complex.
   d. hard to comprehend.

_____11. Biologists use the scientific method to find out about nature. When observations are gathered, the next step is to
   a. design an experiment.
   b. consult a statistician.
   c. formulate a hypothesis.
   d. come up with a theory.

_____12. Valid scientific results should be repeatable by other scientific investigators.
   a. true
   b. false

_____13. A new medication is being tested to determine its effectiveness in controlling obesity in white laboratory mice. A suitable control group would be
   a. any group of mice, caught from the wild.
   b. a group of white laboratory mice, kept in the same way, not given the medication.
   c. a group of white laboratory mice given a lower dosage of the medication.
   d. No control is needed, just see whether the mice lost weight.

_____14. The principle of evolution is a(n)
   a. hypothesis.
   b. unproven observation.
   c. educated guess.
   d. scientific theory.

_____15. For scientific information to be used properly, for the benefit of all living things, we must all be
   a. literate.
   b. smart.
   c. subjective.
   d. socially responsible.

Use the space provided to answer these questions in complete sentences.

16. Why is it preferred to have mathematical data gathered in an experiment?

17. Why is biodiversity important in ecosystems, and why are scientists alarmed at the current rate of loss of biodiversity?

18. Give an example of an area that cannot be subjected to the scientific method.

Test Results: _____ number correct ÷ 18 = _____ × 100 = _____%

ANSWER KEY

STUDY QUESTIONS

1. organized 2. tissues 3. reproduce 4. Humans learn from their parents and others how to act in a socially acceptable, civilized manner, which sometimes sets us apart from the rest of nature. 5. evolution 6. Vertebrates 7. a. plants b. fungi c. protists 8. biosphere 9. population 10. biodiversity 11. extinction 12. a. evolution b. cell c. gene d. homeostasis e. ecosystem 13. a. observation b. hypothesis c. experiment/observations d. conclusion e. scientific theory 14. a. data b. hypothesis c. the control group d. conclusion 15. d 16. e

DEFINITIONS CROSSWORD

Across:
1. biodiversity 3. theory 7. kingdom 8. reproduce 10. evolution 11. hypothesis
Down:
1. biosphere 2. data 4. producer 5. homeostasis 6. vertebrates 9. kingdom 12. cell

CHAPTER TEST

1. d 2. c 3. a 4. b 5. d 6. b 7. c 8. c 9. d 10. a 11. c 12. a 13. b 14. d 15. d 16. Mathematical data can be subjected to statistical analysis; therefore, it is objective and not as likely to be subject to individual bias. 17. Biodiversity helps to ensure that resources are divided up fully and provides structure and stability to ecosystems. The current rapid loss of biodiversity is due to the activities of humans. If loss of biodiversity continues at its present rate, many ecosystems will no longer function as they once did. This could threaten the existence of the biosphere and certainly the survival of humans. 18. Science cannot prove the existence of God, for example.

PART I HUMAN ORGANIZATION

2

CHEMISTRY OF LIFE

It is sometimes difficult for a beginning biology student to appreciate chemistry, but keep in mind that understanding basic chemistry will help you to understand the workings of the cell, the subject of the next chapter.

Take time to fully comprehend the basic structure of an atom. It will help you understand how bonds form between atoms and how large biological molecules are constructed.

Just as there are different uses of the term *center* (in the middle; large building for performing arts; or football position), there are several ways the term *nucleus* is used in biology.

The term *nucleus,* as it is applied to atomic structure (p. 14), is not the same "nucleus" that stores genetic information inside the cell.

You can predict the number of electrons for a given atom by knowing the number of protons in the nucleus. Normally, atoms are electrically neutral, which means the number of positive charges (protons in nucleus) balances the number of negative ones (electrons in shells).

TIP: Be sure to visit the Online Learning Center that accompanies *Human Biology* 9/e. It has practice quizzes, interactive activities, labeling exercises, art quizzes, animations, flash cards, and much more. http://www.mhhe.com/maderhuman9

Study the text section by section. Answer the study questions so that you can fulfill the learning objectives for each section.

2.1 BASIC CHEMISTRY (PAGES 14–17)

After you have answered the questions for this section, you should be able to
- Name and describe the subatomic particles of an atom, indicating which one accounts for the occurrence of isotopes.
- Draw a simplified atomic structure of any atom with an atomic number less than 20.
- Distinguish between ionic and covalent bonds, and draw representative atomic structures for ionic and covalent molecules.

In questions 1–9, fill in the blanks.
1. List three examples of elements in the human body.

   a. _____ b. _____ c. _____

2. The smallest functional unit of an element is a(n) _____.

3. In the following drawing of an atom, write *P* for protons, *N* for neutrons, and *E* for electrons beside *a* and *b*. Indicate the electrical charge for each type of particle.

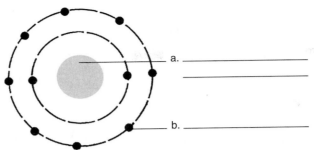

a. _____

_____

b. _____

4. An atom is said to be electrically neutral when it has equal numbers of ᵃ·_____ and ᵇ·_____.

5. Stable, unreactive atoms usually have _____ electrons in their outermost shells.

6. The atomic number of an atom is equal to its number of _____.

7. The atomic mass of an atom is equal to its number of ᵃ·_____ and ᵇ·_____.

8. _____ of an element are atoms that differ in their number of neutrons.

9. Atoms of _____ isotopes, such as carbon 14, radiate high-energy particles from the nucleus as they decay to carbon 12.

In questions 10–16, fill in the blanks.

10. A(n) _____ forms when two or more atoms are bonded together.

11. Sodium (Na) has 11 protons. How many electrons will it have when it is electrically neutral? ᵃ·_____ How many electrons will be in its outermost shell? ᵇ·_____

12. Chlorine (Cl) has 17 protons. How many electrons will chlorine have when it is electrically neutral? ᵃ·_____ How many electrons are in its outermost shell? ᵇ·_____

13. Based on your answers for questions 11 and 12, what type of bond do you expect will form between sodium and chlorine? _____

14. Why? _____

_____

_____

_____

15. When pairs of electrons are shared between atoms, _____ bonds result.

16. In a triple bond, _____ pairs of electrons are shared.

2.2 WATER AND LIVING THINGS (PAGES 18–20)

After you have answered the questions for this section, you should be able to
- Describe the chemical properties of water, and explain their importance for living things.
- Define an acid and base; describe the pH scale, and state the significance of buffers.

In questions 17–23, fill in the blanks.

17. One inorganic molecule, _____, makes up 60–70% of all living things.

18. What type of bond occurs between atoms *within* water molecules? _____

19. What type of bond occurs *between* water molecules? _____

20. Water molecules are a._____ and stick together. Is this property due to hydrogen bonding or to polarity of water molecules? b._____

21. The following diagram illustrates the dissociation of water molecules.

$$H — O — H \rightleftharpoons H^+ + OH^-$$
$$\text{water} \qquad \text{hydrogen} \quad \text{hydroxide}$$
$$\text{ion} \qquad \text{ion}$$

The pH scale is based on the relative abundances of hydrogen ions and hydroxide ions.

When there is more H^+ and less OH^-, is the pH acidic or basic? a._____

When there is less H^+ and more OH^-, is the pH acidic or basic? b._____

22. Technically speaking, a(n) a._____ gives off hydrogen ions in a solution, while a(n) b._____ takes up hydrogen ions or gives off hydroxide ions.

23. A(n) _____ prevents rapid changes in pH by taking up either hydrogen or hydroxide ions.

2.3 MOLECULES OF LIFE (PAGE 22)

After you have answered the questions for this section, you should be able to
- List the four categories of organic molecules.
- Explain how a polymer is formed and how it is broken down.

In questions 24–28, fill in the blanks.

24. In the following diagram, fill in the blanks with the correct subunit or example.

Category	Example	Subunit(s)
Lipids	Fat	a. _____
Carbohydrates	b. _____	Monosaccharide
Proteins	Polypeptide	c. _____
Nucleic acids	d. _____	Nucleotide

25. Are organic molecules characteristic of living or nonliving matter? _____

26. Organic molecules are large in size and are called _____.

27. Macromolecules are built of smaller units, called _____.

28. During digestion, polymers are broken down into component parts by adding a._____ molecules through a reaction referred to as b._____ _____.

2.4 CARBOHYDRATES (PAGES 22–23)

After you have answered the questions for this section, you should be able to
- Give examples of monosaccharides, disaccharides, and polysaccharides, and state their functions.

In questions 29–33, fill in the blanks.

29. The monomer of carbohydrates is a simple sugar, known technically as a(n) a._____. The simple sugar b._____ is one example.

30. Our bodies use glucose as an immediate source of _____.

31. Table sugar, sucrose, is an example of a(n) _____, being built from two sugar molecules.

32. When many glucose molecules are joined, a(n) _____ results.

33. Humans store glucose in the form of a polymer known as a._____ in their livers. Another polysaccharide, b._____, lends structural support to plant cell walls and is not digestible by humans.

2.5 LIPIDS (PAGES 24–25)

After you have answered the questions for this section, you should be able to
- Give examples of various lipids, and state their functions.
- Recognize the structural formula for a saturated and an unsaturated fatty acid.

In questions 34–40, fill in the blanks.

34. Lipids function as a._____ to keep us warm; they provide long-term storage of b._____; and they form a protective c._____ around our internal organs.

35. Dietary fats are most commonly in the form of a._____, which are formed from one b._____ molecule and three c._____ _____ molecules. Since they repel water and are nonpolar, they are called d._____ fats.

36. Emulsifiers break up fats through a process called a._____, in which the emulsifier surrounds individual fat molecules. b._____ from the gallbladder functions in the same way in the digestive system.

37. Write the word *saturated* or *unsaturated* on the line below the appropriate structure.

a. _____   b. _____

38. What does saturated mean? _____
_____

39. A type of lipid with a polar phosphate group, called a(n) a._____, is a major component of plasma membranes.

40. Cholesterol, a(n) _____, is a precursor to sex hormones in the body.

2.6 PROTEINS (PAGES 26–27)

After you have answered the questions for this section, you should be able to
- Give examples of proteins, and state their functions.
- Recognize an amino acid and demonstrate how a peptide bond is formed.

In questions 41–45, fill in the blanks.

41. Certain proteins can function as a._____ to catalyze chemical reactions in the cell. Fingernails contain another important protein, called b._____. Another protein, called c._____, is an important structural component of connective tissue.

42. The monomer of proteins is called a(n) _____.

43. In the following diagram of an amino acid, —NH_2 is the a._____ group.
—COOH is the b._____ group.

$$H_2N-\overset{\overset{H}{|}}{\underset{\underset{R}{|}}{C}}-COOH$$

c. How does a peptide bond form between two adjacent amino acids?
_____
_____
_____

44. a._____ structure in proteins refers to the sequence of amino acids. b._____ structure comes about when the polypeptide takes on a certain orientation in space. The c._____ structure of a protein is its final three-dimensional shape. Polypeptides are sometimes arranged to give the d._____ level of protein structure.

45. When proteins are exposed to heat, chemicals, or extremes in pH, they undergo a change in shape. They no longer function as usual, so we say they have been _____.

2.7 NUCLEIC ACIDS (PAGES 29–30)

After you have answered the questions for this section, you should be able to
- Give examples of nucleic acids, and state their functions.
- State the components of a nucleotide, and tell how these monomers are joined to form a nucleic acid.
- Compare the structures of DNA and RNA.
- Describe the structure and function of ATP.

In questions 46–49, fill in the blanks.

46. The monomers of nucleic acids are _____.

47. The nucleic acid a._____ stores the genetic information for the cell. Another nucleic acid, b._____, works with DNA to specify the order of amino acids during the synthesis of proteins.

48. Refer to the following diagram of a strand of nucleotides to answer questions a–d.

   a. What molecule is represented by S? _____
   b. What molecule is represented by B? _____
   c. How many different types of B are in DNA? _____
   d. What type of bond joins the two strands of DNA? _____

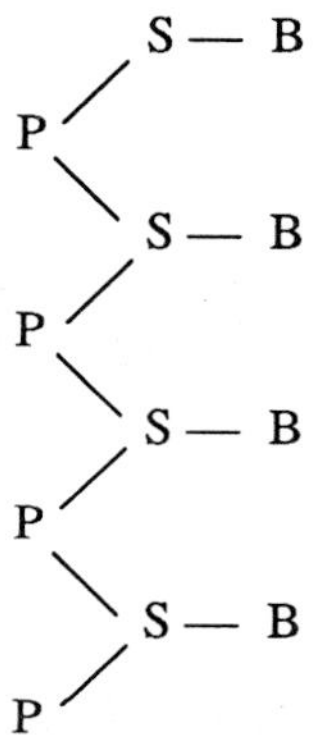

```
         S — B
      P /
        \
         S — B
      P /
        \
         S — B
      P /
        \
         S — B
      P /
```

49. One nucleotide, a._____, functions as an energy carrier inside cells. Energy is released from this molecule when a(n) b._____ group is removed.

50. Complete this summary table.

	Building Blocks	Function
Polysaccharides		
Proteins		

Lipids
  Fats
  Phospholipids		

Nucleic acids
  DNA
  RNA		

Questions 51–52 pertain to the following diagram.

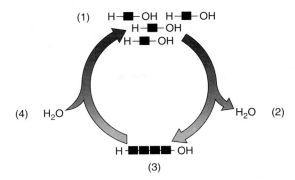

51. a. What building blocks should be associated with (1) in the diagram?

   _____

   b. What polymers should be associated with (3)? _____

   _____

   c. Which number in the diagram stands for a hydrolysis reaction? _____

   d. Which number in the diagram stands for a dehydration reaction? _____

52. Of the molecules listed in the first column of the table in question 50,

   a. Which are *most* concerned with energy? _____

   b. Which one forms enzymes? _____

   c. Which one makes up genes? _____

DEFINITIONS CROSSWORD

Review key terms by completing this crossword puzzle, using the following alphabetized list of terms:

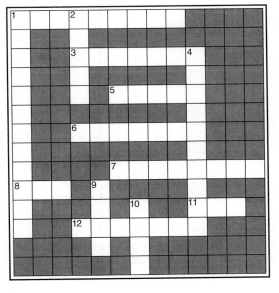

atom
ATP
base
carbohydrate
cellulose
DNA
electron
glucose
ionic
isotope
lipid
nucleotide
protein

Across

1 Complex polysaccharide found in plant tissue that we cannot digest.
3 Organic compound made up of long chains of amino acids.
5 Type of bond formed between oppositely charged ions.
6 Monosaccharide that serves as a building block for starch.

7 A negatively charged subatomic particle.
8 Nucleotide that serves as the energy carrier for the cell.
11 A nucleic acid that serves as the cell's genetic material.
12 An atom that differs in its number of neutrons.

Down

1 Organic compound type that includes mono- and polysaccharides.
2 Insoluble organic compound, such as fats.
4 Building block of nucleic acid.
9 Molecule that releases OH⁻ in solution.
10 Functional unit of an element.

11

OBJECTIVE QUESTIONS

Do not refer to the text when taking this test.

____ 1. The atomic weight of an atom is determined by the number of
   a. protons.
   b. neutrons.
   c. electrons.
   d. protons and neutrons.

____ 2. Isotopes differ due to the number of
   a. protons.
   b. neutrons.
   c. electrons.
   d. protons and neutrons.

____ 3. When an atom either gains or loses an electron, it becomes
   a. electrically neutral.
   b. an ion.
   c. stable.
   d. unreactive.

____ 4. Chlorine has 17 protons. When chlorine becomes the chloride ion (Cl^-), it has
   a. gained an electron.
   b. lost an electron.
   c. gained a proton.
   d. lost a neutron.

____ 5. When sodium interacts with chlorine, sodium loses an electron, while chlorine gains one. This interaction forms
   a. an ionic bond.
   b. a condensation synthesis.
   c. a condensation.
   d. a covalent bond.

____ 6. Bonds between carbon and oxygen are generally
   a. hydrogen bonds.
   b. ionic bonds.
   c. covalent bonds.
   d. weak and highly transient.

____ 7. When a strong acid, such as hydrochloric acid, is added to water,
   a. hydrogen ions are taken up.
   b. hydroxide ions are released.
   c. hydrogen ions are released.
   d. the pH stays the same.

____ 8. A pH of 11.5 is considered to be
   a. slightly acidic.
   b. strongly acidic.
   c. strongly basic.
   d. about neutral.

____ 9. A buffer is designed to
   a. increase the concentration of H^+.
   b. increase the concentration of OH^-.
   c. take up excess H^+ or OH^-.
   d. increase the concentration of salt.

____ 10. Polymers are
   a. chains of building-block molecules.
   b. formed by dehydration.
   c. broken down by hydrolysis.
   d. All of these are correct.

____ 11. Which pair of molecules is mismatched?
   a. amino acid—protein
   b. fatty acid—lipid
   c. glucose—starch
   d. glycerol—nucleic acid

For questions 12–16, match the statements to these terms.
   a. triglyceride   b. unsaturated fatty acid
   c. saturated fatty acid   d. phospholipid

____ 12. Fat or oil.

____ 13. Has one or more double bonds along the fatty acid chain.

____ 14. Made up of a glycerol and three fatty acid molecules.

____ 15. Has hydrogen at every position along the fatty acid chain.

____ 16. A major component of plasma membranes.

____ 17. Of the following structures, which molecule is unsaturated?

____ 18. The primary structure of a protein refers to its
   a. three-dimensional shape.
   b. order of amino acids.
   c. order of nucleic acids.
   d. orientation of the amino acids.

____ 19. Proteins are polymers of _____, which sometimes function to _____.
   a. amino acids; catalyze chemical reactions
   b. glucose; build muscle strength
   c. nucleotides; synthesize proteins
   d. ribosomes; produce quick energy

_____20. In any amino acid, —NH$_2$ is the _____ group, and —COOH is the _____ group.
- a. carboxyl; amino
- b. amino; carboxyl
- c. peptide; fatty acid
- d. hydroxyl; carboxyl

_____21. A bond forming between the —NH$_2$ group of one amino acid and the —COOH group of another amino acid is called a(n)
- a. double bond.
- b. ionic bond.
- c. hydrogen bond.
- d. peptide bond.

_____22. Twisting of a chain of amino acids into an α-helix is termed the _____ structure of a protein.
- a. primary
- b. secondary
- c. tertiary
- d. quaternary

In questions 23–26, match the statements to these terms:
- a. RNA b. DNA c. nucleotide d. ATP

_____23. Building block of the nucleic acids.
_____24. Functions with DNA to specify the order of amino acids in a protein.
_____25. The genetic material of the cell.
_____26. The energy currency of the cell.

THOUGHT QUESTIONS

Answer in complete sentences.

27. Give one example of how radioactive isotopes are used in research or medicine.

28. Describe two ways in which the properties of water allow the existence of life.

Test Results: _____ number correct ÷ 28 = _____ × 100 = _____%

ANSWER KEY

STUDY QUESTIONS

1. a. carbon **b.** hydrogen **c.** oxygen (or nitrogen) **2.** atom **3. a.** protons (positive charge), neutrons (no charge) **b.** electrons (negative charge) **4. a.** protons **b.** electrons **5.** eight **6.** protons **7. a.** protons **b.** neutrons **8.** Isotopes **9.** radioactive **10.** molecule **11. a.** eleven **b.** one **12. a.** seventeen **b.** seven **13.** ionic **14.** Once sodium gives up the electron in its outermost shell, it is a stable atom. Chlorine also becomes stable by accepting the electron from sodium, now with eight electrons in its outermost shell. Since the chlorine atom now bears a net negative charge, and sodium a net positive charge, the two atoms (now ions) are attracted to each other and held together by an ionic bond. **15.** covalent **16.** three **17.** water **18.** polar covalent bond **19.** hydrogen bond **20. a.** cohesive **b.** hydrogen bonding **21. a.** acidic **b.** basic **22. a.** acid **b.** base **23.** buffer **24. a.** glycerol and fatty acids **b.** polysaccharide **c.** amino acid **d.** DNA, RNA **25.** living **26.** macromolecules **27.** subunits **28. a.** water **b.** hydrolysis reaction **29. a.** monosaccharide **b.** glucose **30.** energy **31.** disaccharide **32.** polysaccharide **33. a.** glycogen **b.** cellulose **34. a.** insulation **b.** energy **c.** padding **35. a.** triglycerides **b.** glycerol **c.** fatty acid **d.** neutral **36. a.** emulsification **b.** Bile **37. a.** unsaturated **b.** saturated **38.** The carbon chain has all the hydrogens it can hold; there are no double bonds between carbon atoms. **39.** phospholipid **40.** steroid **41. a.** enzymes **b.** keratin **c.** collagen **42.** amino acid **43. a.** amino **b.** carboxyl **c.** When the carboxyl (acidic) group of one amino acid reacts with the amino group of another amino acid, a molecule of water is removed, and the peptide bond forms. **44. a.** Primary **b.** Secondary **c.** tertiary **d.** quaternary **45.** denatured **46.** nucleotides **47. a.** DNA **b.** RNA **48. a.** sugar **b.** base **c.** four **d.** hydrogen **49. a.** ATP **b.** phosphate

50.

	Building Blocks	Function
Polysaccharides	Monosaccharides	Energy storage
Proteins	Amino acids	Enzymes speed chemical reactions; structural components (e.g., muscle proteins)
Lipids Fats Phospholipids	Glycerol, 3 fatty acids Glycerol, 2 fatty acids, phosphate group	Long-term energy storage Plasma membrane structure
Nucleic acids DNA RNA	Nucleotide with deoxyribose sugar Nucleotide with ribose sugar	Genetic material Protein synthesis

51. a. glucose, amino acids, glycerol and fatty acids, nucleotides, phosphate group **b.** fats, polysaccharide, proteins, phospholipid, DNA, RNA **c.** 4 **d.** 2
52. a. polysaccharides and lipids **b.** protein **c.** DNA

DEFINITIONS CROSSWORD

Across:
1. cellulose **3.** protein **5.** ionic **6.** glucose **7.** electron **8.** ATP **11.** DNA **12.** isotope

Down:
1. carbohydrate **2.** lipid **4.** nucleotide **9.** base **10.** atom

CHAPTER TEST

1. d **2.** b **3.** b **4.** a **5.** a **6.** c **7.** c **8.** c **9.** c
10. d **11.** d **12.** a **13.** b **14.** a **15.** c **16.** d **17.** b
18. b **19.** a **20.** b **21.** d **22.** b **23.** c **24.** a **25.** b
26. d **27.** Radioactive isotopes are used to determine the age of fossils. They are also used in many diagnostic medical tests. **28.** Because ice is less dense than water, bodies of water freeze from the top down. Aquatic organisms can survive throughout the winter in the cold water beneath the ice. The cohesiveness of water enables water to travel easily through a tube, as blood, a liquid, does within blood vessels.

3

CELL STRUCTURE AND FUNCTION

STUDY QUESTIONS

Study the text section by section. Answer the study questions so that you can fulfill the learning objectives for each section.

3.1 WHAT IS A CELL (PAGES 36–37)

After you have answered the questions for this section, you should be able to
- Explain why cells are small by nature.
- Recognize that different types of microscopy reveal different features of the cell.

1. Cells are small to enable them to maintain a favorable ᵃ._____ to ᵇ._____ ratio. This helps the diffusion of nutrients and gases throughout the cell.

2. As a student, you are most likely to encounter the ᵃ._____ microscope in the laboratory. The ᵇ._____ electron microscope scans the surfaces of samples to look at surface features. The ᶜ._____ electron microscope is capable of viewing the internal features of cells. Which type of microscope has the best resolving power? ᵈ._____

3.2 CELLULAR ORGANIZATION (PAGES 38–47)

After you have answered the questions for this section, you should be able to
- Describe the structure of the plasma membrane.
- Explain the variety of methods used by cells to allow passage of materials through the plasma membrane.
- Predict the results of placing a cell in solutions of various tonicity.
- Describe the structure of the cytoskeleton within the cytoplasm.
- Describe the structure of the nucleus and its importance to the cell.
- Discuss the features and importance of ribosomes.
- Explain the relationship among the endoplasmic reticulum, the Golgi apparatus, and lysosomes.
- Describe the function of cilia and flagella.
- Explain the structure and function of the mitochondria within the cell.

3. Label the parts of the plasma membrane in the following diagram using the alphabetized list of terms.

carbohydrate chain
cholesterol
filaments of cytoskeleton
glycolipid
glycoprotein
phospholipid bilayer
protein

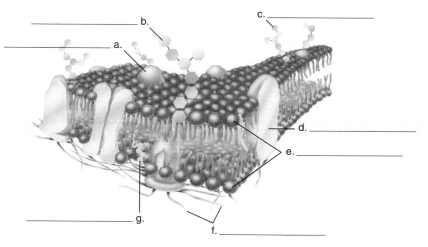

a. _____

b. _____

c. _____

d. _____

e. _____

f. _____

g. _____

4. Within the bilayer, a._____, or hydrophilic, heads extend outward and nonpolar, or b._____, fatty acid tails extend to the interior of the membrane.

5. The plasma membrane is selectively permeable because it _____ what substances cross through the membrane.

6. a._____ is the passive movement of molecules from an area of greater concentration to an area of lower concentration. b._____ is a special case of the same process, involving the movement of water across a membrane.

7. If a cell is placed in a(n) a._____ environment, it neither gains nor loses water. When a cell enters a(n) b._____ solution, it gains water until it bursts.

8. a._____ transport involves a carrier protein assisting a larger molecule across the membrane with no expenditure of energy. b._____ transport carries materials across membranes against a concentration gradient and requires energy.

9. During the process known as a._____, the plasma membrane pinches off a vesicle around a particle the cell is engulfing. b._____ is the reverse process, during which the cell expels contents to the outside.

10. Label the cellular structures in the following diagram using the alphabetized list of terms.

actin filament
centriole
chromatin
cytoplasm
Golgi apparatus
lysosome
microtubule
mitochondrion
nuclear envelope
nuclear pore

nucleolus
nucleus
plasma membrane
polyribosome
ribosomes
rough ER
smooth ER
vacuole
vesicle

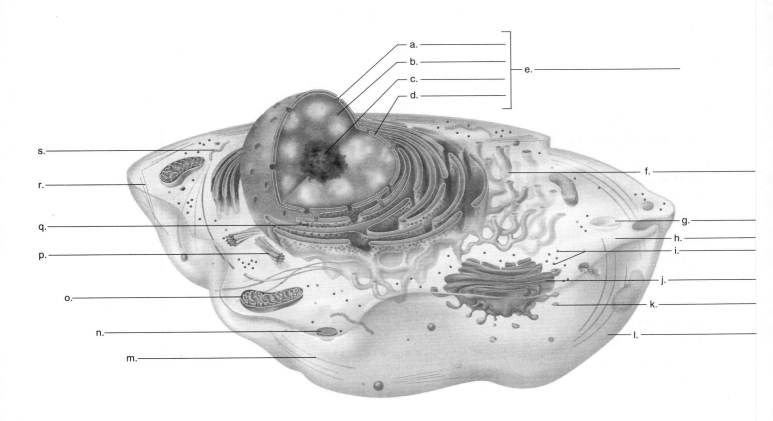

11. Match these functions with the cell parts listed in question 10. Not all cell parts will be needed.
    a. _____ lipid metabolism
    b. _____ contains digestive enzymes
    c. _____ site of protein synthesis
    d. _____ location of the nucleolus
    e. _____ site of ATP production
    f. _____ give rise to basal bodies
    g. _____ processes, packages, and distributes molecules
    h. _____ found in flagella and cilia
    i. _____ studded with ribosomes
    j. _____ storage of substances
    k. _____ ribosomal production

12. The nucleus is enclosed by the a._____, which contains b._____ through which materials can pass into the cytoplasm. Within the nucleus, the c._____ is the site of production of rRNA for the manufacture of d._____. Ribosomes are associated with e._____ endoplasmic reticulum, where proteins are made. Proteins travel in vesicles to the f._____, which packages them for secretion.

13. The cytoskeleton has two types of filaments of interest to us, a._____ and b._____. Centrioles give rise to c._____, which direct the formation of d._____ and e._____.

14. In 14a and b, label the following diagram of a mitochondrion using these terms: matrix, cristae.

a.   b.

3.3 CELLULAR METABOLISM (PAGES 48–50)

After you have answered the questions for this section, you should be able to
• Understand the functioning of a metabolic pathway and the importance of enzymes within it.
• Describe a generalized equation for an enzymatic reaction.
• Explain the function of coenzymes.
• Understand the activities that occur during cellular respiration, and relate the results to the three metabolic pathways involved.
• Associate mitochondria with the citric acid cycle and the electron transport system.
• Understand what conditions are needed for the process of fermentation compared to those of cellular respiration.

15. A a._____ _____ is a series of b._____-catalyzed reactions that must occur in a particular sequence.

16. The interaction of an enzyme with its substrate can be summarized as follows: E + S —> a._____ —> E + P, where E stands for b._____, S for c._____, and P for d._____.

17. Nonprotein organic molecules that sometimes assist the functioning of enzymes are called a._____.
    b._____ is one example.

18. Which of the terms in question 14 do you associate with the citric acid cycle? a._____ Which of the terms do you associate with the electron transport system? b._____ Where does glycolysis take place?
    c._____

19. Complete the following diagram by labeling the pathways and by adding where needed: ATP, CO_2, O_2, and H_2O.

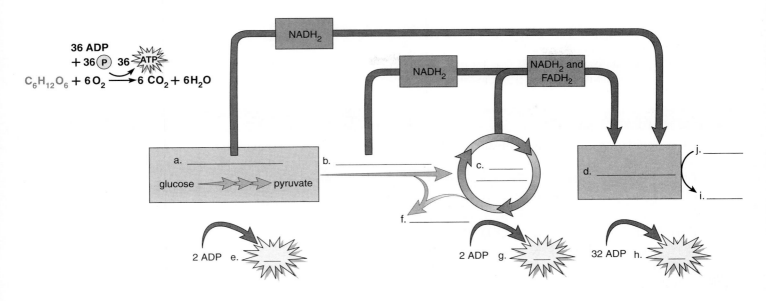

In 20a, use the space provided to answer this question using a complete sentence.

20. Why do arrows point from each of the pathways in the diagram to the electron transport system? a._____

_____

During cellular respiration, glucose is b._____ to c._____ and water. As breakdown occurs

during glycolysis, the preparatory reaction, and the citric acid cycle, hydrogen atoms ($H^+ + e^-$) are removed and carried

to the d._____ in the mitochondria. As the electrons are passed from carrier to carrier, energy is released

and 32 molecules of e._____ are produced.

21. Match the definitions to the pathways you labeled in question 19:
    a. _____ A series of reactions that begins with glucose and ends with pyruvate with the buildup of 2 ATP molecules.
    b. _____ A circular series of reactions in the mitochondria that receives acetyl groups and metabolizes them to carbon dioxide with a net gain of 2 ATP.
    c. _____ Pyruvate is converted to an acetyl group and carbon dioxide is released.
    d. _____ A chain or carrier in mitochondria; begins with $NADH_2$ and ends when oxygen is reduced to water.

In 22–24, use the following equation:

a.	b.	c.	d.	e.
$C_6H_{12}O_6$ +	$6O_2$ ⟶	$6CO_2$ +	$6H_2O$ +	energy

22. Write the name of each molecule or type of energy being produced in the equation:
    a. _____
    b. _____
    c. _____
    d. _____
    e. _____

23. This is an overall equation for _____.

24. Match the following pathways to the names of the molecules in the equation:
    a. _____ , _____ Glycolysis
    b. _____ Preparatory reaction
    c. _____ , _____ Citric acid cycle
    d. _____ , _____ , _____ Electron transport chain

25. Indicate whether these statements about fermentation are true (T) or false (F).

Fermentation:
a. _____ is an anaerobic process.
b. _____ produces as much ATP as cellular respiration.
c. _____ is unique to yeast.
d. _____ results in lactate buildup.
e. _____ results in only 2 ATP.
f. _____ occurs when oxygen is not available.

DEFINITIONS WORDSEARCH

Rewrite key terms by completing this wordsearch, using the following alphabetized list of terms:

```
S E E T H E C A T R U N T T H I M O
E M O S O M O R H C B R P F A M I P
T H E R I M O T R Y C Y L T T O C K
T H E R I O T B G T D O A G D H R E
G I N N Y O C Y T O S K E L E T O N
R E N T O F R D C P T Y I O S H T I
A Y L I L L K A Q L W Z E U P P U P
F U N I D A R Y T A Y V L K C I B L
R E T N U G G U N S L O O H G F U M
L I T C O E N Z Y M E U I U O D L J
K I L Y T L N U T L R U R O M M E E
F L A G G L T M C I Y O T R E E S I
G O L G A U B U S O K J N O P L Y G
C I L I U M N R T E L L E N A G R O
N U C L A I U B F I T O C J U M P G
```

chromosome
cilium
coenzyme
cytoplasm
cytoskeleton
flagellum
microtubule
nucleolus
organelle

a. _____ Rodlike structure in the nucleus; DNA.
b. _____ Short, hairlike extension from the cell.
c. _____ Slender, long extension from the cell.
d. _____ Structure within the cell, surrounded by a membrane, with a specific function.
e. _____ Region inside the nucleus where rRNA is produced for the manufacture of ribosomes.
f. _____ Cytoskeleton component made up of the protein tubulin.
g. _____ Nonprotein organic molecule that aids the action of an enzyme.
h. _____ Network of protein filaments found within the cell. Its function is to move organelles around, among other things.
i. _____ Semifluid medium outside the nucleus in the cell.

CHAPTER TEST

OBJECTIVE TEST

Do not refer to the text when taking this test.

_____ 1. Cells are small so that
   a. they require fewer nutrients.
   b. they can reproduce rapidly.
   c. materials can easily diffuse throughout the cell.
   d. they can more readily become components of tissues.

_____ 2. Which type of microscopy has the best resolving power?
   a. simple compound microscope
   b. scanning electron microscope
   c. transmission electron microscope
   d. Both electron microscopes have the same resolving power.

_____ 3. Which type of molecule forms a bilayer within the plasma membrane?
   a. carbohydrate
   b. protein
   c. phospholipid
   d. nucleic acid

_____ 4. A small, lipid-soluble molecule readily passes through the plasma membrane. Which is the most likely explanation?
   a. A carrier protein must be carrying out facilitated transport.
   b. Diffusion occurs, because the membrane is made up of lipid molecules.
   c. The cell is expending energy to do work.
   d. Vacuole formation occurs as the membrane engulfs a particle.

_____ 5. Which of these does not require an expenditure of energy?
   a. diffusion
   b. osmosis
   c. facilitated transport
   d. None of these require energy.

_____ 6. A carrier is moving a substance from an area of greater concentration outside to one of lower concentration inside the cell. Which of these is occurring?
   a. facilitated transport
   b. diffusion
   c. active transport
   d. vacuole formation

_____ 7. The thyroid gland contains a higher concentration of iodine than the blood of the body. This is most likely due to
   a. diffusion.
   b. active transport.
   c. facilitated transport.
   d. endocytosis.

_____ 8. An animal cell always takes in water when it is placed in a(n)_____ solution.
   a. hypertonic
   b. osmotic
   c. isotonic
   d. hypotonic

_____ 9. The nucleus is considered the control center of the cell because it
   a. is located in the exact center of the cell.
   b. houses the nucleolus.
   c. contains the genetic information of the cell.
   d. has nuclear pores through which materials can pass.

_____10. Which of these cellular structures is NOT composed of membranes?
   a. ribosomes
   b. Golgi apparatus
   c. lysosomes
   d. endoplasmic reticulum

_____11. Which organelle has a 9 + 0 arrangement of microtubules and is associated with the formation of basal bodies?
   a. actin filaments
   b. spindle fibers
   c. flagellum
   d. centriole

_____12. Cilia and flagella contain
   a. centrioles.
   b. microtubules.
   c. basal bodies.
   d. actin filaments.

_____13. The cytoskeleton of the cell is composed of
   a. protein.
   b. actin filaments.
   c. microtubules.
   d. All of these are correct.

In questions 14–19, match the functions to the cellular structure.
   a. mitochondria   b. ribosomes   c. lysosomes
   d. centrioles   e. Golgi apparatus   f. cilia

_____14. packaging and secretion
_____15. hydrolytic digestive enzymes
_____16. energy production
_____17. protein synthesis
_____18. cell division
_____19. cell locomotion

_____20. NAD performs what function in cellular respiration?
   a. It is an oxygen carrier.
   b. It carries hydrogen atoms.
   c. It functions as an enzyme.
   d. It facilitates the preparatory reaction.

_____21. In a metabolic pathway, each enzyme catalyzes a specific reaction in
   a. sequence.
   b. random order.
   c. organelles only.
   d. anaerobic conditions.

_____22. The pathway responsible for most of the ATP formed during cellular respiration is
   a. glycolysis.
   b. the citric acid cycle.
   c. the preparatory reaction.
   d. the electron transport chain.

_____23. The pathway that begins with glucose and ends with pyruvate is _____, and it occurs in the _____.
   a. glycolysis; cytoplasm
   b. the citric acid cycle; mitochondria
   c. the transition reaction; cytoplasm
   d. glycolysis; cristae of the mitochondria

_____24. The electron transport chain is located where?
   a. within the matrix of the mitochondria
   b. on the rough endoplasmic reticulum
   c. on the ribosomes
   d. on the cristae of the mitochondria

_____25. The final acceptor for hydrogen in cellular respiration is
   a. ATP.
   b. NAD.
   c. FAD.
   d. oxygen.

THOUGHT QUESTIONS

Answer in complete sentences.
   26. What happens at the end of glycolysis when no oxygen is present in the cell?

   27. If a cell is placed into a hypertonic environment, what is the result?

Test Results: _____ number correct ÷ 27 = _____ × 100 = _____%

ANSWER KEY

STUDY QUESTIONS

1. a. surface area **b.** volume   **2. a.** simple compound **b.** scanning **c.** transmission **d.** either electron microscope **3. a.** glycoprotein **b.** carbohydrate chain **c.** glycolipid **d.** protein **e.** phospholipid bilayer **f.** filaments of cytoskeleton **g.** cholesterol   **4. a.** polar **b.** hydrophobic **5.** regulates   **6. a.** Diffusion **b.** Osmosis   **7. a.** isotonic **b.** hypotonic   **8. a.** Facilitated **b.** Active   **9. a.** endocytosis **b.** Exocytosis   **10. a.** nuclear pore **b.** chromatin **c.** nucleolus **d.** nuclear envelope **e.** nucleus **f.** smooth ER **g.** vacuole **h.** cytoplasm **i.** ribosomes **j.** Golgi apparatus **k.** vesicle **l.** plasma membrane **m.** microtubule **n.** lysosome **o.** mitochondrion **p.** centriole **q.** rough ER **r.** actin filaments **s.** polyribosome   **11. a.** smooth ER **b.** lysosome **c.** ribosomes **d.** nucleus **e.** mitochondrion **f.** centriole **g.** Golgi apparatus **h.** microtubules **i.** rough ER **j.** vacuole **k.** nucleolus   **12. a.** nuclear envelope **b.** pores **c.** nucleolus **d.** ribosomes **e.** rough **f.** Golgi apparatus   **13. a.** microtubules **b.** actin filaments **c.** basal bodies **d.** cilia **e.** flagella   **14. a.** cristae **b.** matrix   **15. a.** metabolic pathway **b.** enzyme   **16. a.** ES **b.** enzyme **c.** substrate **d.** product **17. a.** coenzymes **b.** NAD   **18. a.** matrix **b.** cristae **c.** in the cytoplasm outside the mitochondrion   **19. a.** glycolysis **b.** preparatory reaction **c.** citric acid cycle **d.** electron transport chain **e.** ATP **f.** CO_2 **g.** ATP **h.** ATP **i.** H_2O **j.** O_2 **20. a.** As breakdown occurs during glycolysis, the preparatory reaction, and the citric acid cycle, hydrogen atoms are removed from substrates and carried to the electron transport chain. **b.** broken down **c.** carbon dioxide **d.** electron transport chain **e.** ATP   **21. a.** a **b.** c **c.** b **d.** d   **22. a.** glucose **b.** oxygen **c.** carbon dioxide **d.** water **e.** ATP   **23.** cellular respiration   **24. a.** glucose, ATP **b.** carbon dioxide **c.** carbon dioxide, ATP **d.** oxygen, water, ATP   **25. a.** T **b.** F **c.** F **d.** T **e.** T **f.** T

DEFINITIONS WORDSEARCH

a. chromosome **b.** cilium **c.** flagellum **d.** organelle **e.** nucleolus **f.** microtubule **g.** coenzyme **h.** cytoskeleton **i.** cytoplasm

```
                                          M
     E M O S O M O R H C               I
                     Y                 C
                     T                 R
            C Y T O S K E L E T O N
       F           P       S       T
       L           L     U         U
       A           A   L           B
       G           S O             U
     C O E N Z Y M E               L
       L           L               E
       L       C
       U   U
     C I L I U M N     E L L E N A G R O
```

CHAPTER TEST

1. c   **2.** c   **3.** c   **4.** b   **5.** d   **6.** a   **7.** b   **8.** d   **9.** c **10.** a   **11.** d   **12.** b   **13.** d   **14.** e   **15.** c   **16.** a **17.** b   **18.** d   **19.** f   **20.** b   **21.** a   **22.** d   **23.** a   **24.** d **25.** d   **26.** When no oxygen is present, the pyruvate cannot enter the preparatory reaction and be transported into the mitochondria to engage in the citric acid cycle. Instead, the cell turns to fermentation pathways to release a small quantity of energy. The result is the buildup of lactic acid within the cell.   **27.** A cell placed in a hypertonic environment will shrink as it loses water to its surroundings.

4

ORGANIZATION AND REGULATION OF BODY SYSTEMS

In the chapters that follow, the various organ systems are discussed. Having a good understanding of tissues will help you to appreciate the complexity of organs and organ systems. Homeostasis is a recurring theme throughout the remaining chapters. Many homeostatic mechanisms operate by negative feedback, much as a thermostat regulates the temperature of a room. Keep this in mind as an example of how homeostasis works as you study upcoming chapters.

Biological terminology is often confusing, especially to a student new to the discipline. In this chapter, you will note the use of the term *fibers* as reinforcing protein strands within connective tissue (p. 56). *Fibers* also refers to muscle cells. Be aware that it is easy to confuse the two. Another example of a term having more than one definition is the use of *membranes* to indicate structures within or surrounding the cell and the use of *membranes* in this chapter in reference to body linings (p. 68). Be sure to determine the context in which the term *membrane* is used.

TIP: Be sure to visit the Online Learning Center that accompanies *Human Biology* 9/e. It has practice quizzes, interactive activities, labeling exercises, art quizzes, animation, flash cards, and much more. http://www.mhhe.com/maderhuman9

STUDY QUESTIONS

Study the text section by section. Answer the study questions so that you can fulfill the learning objectives for each section.

4.1 TYPES OF TISSUES (PAGES 56–63)

After you have answered the questions for this section, you should be able to
- Discuss the different types of connective tissues and their functions.
- Describe the types of muscle tissue and how they operate.
- State the characteristics and functions of nervous tissue.
- Describe the features and functions of epithelial tissues.
- Explain how junctions between cells aid communication between them.

1. Loose, fibrous connective tissue has cells called a._____, plus fibers made from b._____ and c._____. It binds the skin to underlying organs.

2. Tendons and ligaments are made up of a._____ tissue. Tendons join b._____ to bone, and ligaments join c._____ to bone.

3. a._____ cartilage can be found at the ends of bones and makes up the fetal skeleton. In it, cells lie within b._____, surrounded by a gel-like c._____.

4. The type of cartilage found in the ear, called a._____ cartilage, has many b._____ fibers to add flexibility. c._____, found in the intervertebral disks, aids in cushioning against jolts.

5. Draw a sketch of bone tissue. Label these parts: lacunae, matrix, and osteons (Haversian systems).

6. a._____ blood cells carry oxygen throughout the body in association with molecules of hemoglobin.
   b._____ blood cells are responsible for immunity, and cell fragments called c._____ aid in blood clotting.

7. Complete this table.

	Fiber Appearance	Location	Control
Skeletal	a.		
Smooth	b.		
Cardiac	c.		

8. The brain and spinal cord are made up of cells called a._____. Outside the central nervous system, connective tissue binds the long fibers of these cells to form b._____. The function of a neuron is to c._____. The other types of cells in nervous tissue are the d._____. These types of cells provide e._____ to neurons and keep tissue free of debris. Neuroglia outnumber neurons f._____ to one and take up more than g._____ the volume of the brain.

9. Epithelium can be classified by a._____ . b._____ epithelium has cube-shaped cells, while c._____ epithelium has elongated, cylindrical cells.

10. Draw a diagram of squamous epithelium.
    a.

    b. Name one place in the human body where squamous epithelium can be found. _____
    c. What is the function of this tissue? _____
    d. When this tissue is found in layers, it is called _____ squamous epithelium.

11. The windpipe is lined by pseudostratified ciliated columnar epithelium. Use the space provided to describe this tissue in complete sentences. _____
_____
_____
_____

12. a. _____ junctions allow for materials and information to be exchanged between cells.
    b. _____ junctions allow cells to stretch and bend.
    c. _____ junctions bind cells firmly together and form an impermeable barrier.

After you have answered the questions for this section, you should be able to
• List the three regions of the skin, their functions, and the structures located within them.

13. Skin has an outer layer called the ᵃ._____. The cells of this layer become waterproof once they are filled

with the protein ᵇ._____.

14. An accessory organ of the skin, the ᵃ._____, forms hair shafts. ᵇ._____ muscles cause goose

bumps to appear. ᶜ._____ glands provide sebum to lubricate hairs and skin.

15. Label the following diagram of the skin using the alphabetized list of terms.
    adipose tissue
    arrector pili muscle
    blood vessels
    dermis
    epidermis
    hair follicle
    hair root
    hair shaft
    nerve
    oil gland
    sensory receptors
    subcutaneous layer
    sweat gland

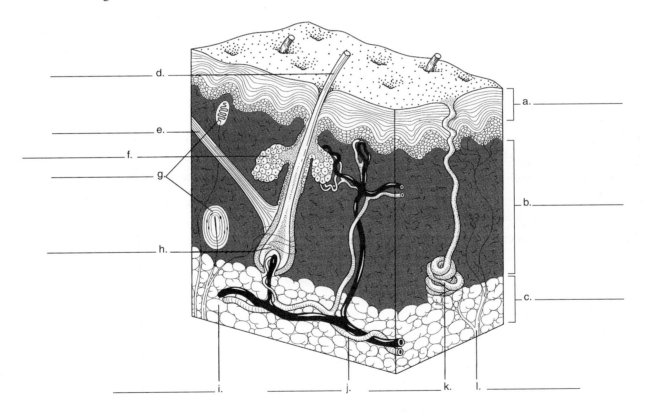

16. Why does sweat lower body temperature? _____

_____

25

After you have answered the questions for this section, you should be able to
- Give examples of the various organ systems of the body and their general functions.
- Describe the body cavities and the organs they contain.
- Describe the body membranes.

17. Place the appropriate letter next to each system.

    M—maintenance of the body   S—support and movement

    R—regulation of body systems   C—continuance of the species

    a. _____ digestive system

    b. _____ reproductive system

    c. _____ skeletal system

    d. _____ nervous system

    e. _____ respiratory system

18. The a._____ system is responsible for reducing food to molecules that can be used by the cells of the body.

    The b._____ system allows us to move from one place to another and generates heat. The

    c._____ system consists of the brain, spinal cord, nerves, and sensory receptors. The d._____

    system sends out hormones to regulate bodily functions.

19. Specify the correct cavity for each organ, using these letters:

    T—thoracic cavity   A—abdominal cavity   D—dorsal cavity

    a. _____ small intestine   e. _____ lungs

    b. _____ brain   f. _____ stomach

    c. _____ bladder   g. _____ liver

    d. _____ heart   h. _____ kidneys

20. A large, ventral cavity called a(n) a._____ can be seen during embryological development. It later

    develops into the b._____ cavity of the chest and the c._____ cavity of the lower abdomen.

21. The a._____ are connective tissue membranes protecting the brain and spinal cord. b._____

    lines the gut and secretes mucus. c._____ membrane lines joint cavities and secretes lubricating

    d._____.

22. Pleura and peritoneum are examples of a._____ membranes. These secrete b._____ to keep the

    membranes moist and to prevent sticking.

After you have answered the questions for this section, you should be able to
- Define homeostasis, tell how it is controlled, and explain its importance.
- Understand how the nervous and endocrine systems coordinate to maintain homeostasis.
- Use the maintenance of room temperature as an example of negative feedback control.

Use the space provided to answer these questions in complete sentences.

23. a. What is homeostasis? _____

    b. Give an example of homeostasis. _____

24. Label the adjacent diagram of a negative feedback mechanism, using these terms:

    effect   control center   homeostasis
    sensor   stimulus

25. Give an example of negative feedback control in the body. Use the space provided to give your example in complete sentences. _____

    _____
    _____
    _____
    _____

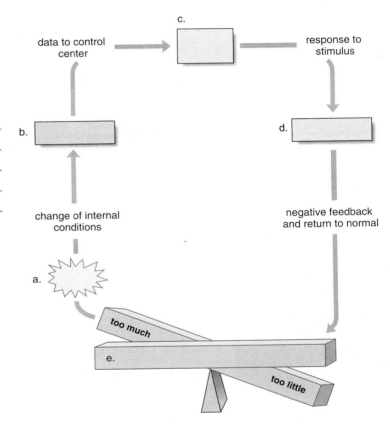

26. Indicate whether these statements about positive feedback are true (T) or false (F).

    a. _____ Like negative feedback, positive feedback maintains homeostasis.

    b. _____ Brings about change in the same direction.

    c. _____ Occurs during childbirth.

    d. _____ Occurs when a thermostat regulates room temperature.

In questions 27–28, fill in the blanks.

27. How does each of the following systems contribute to homeostasis?

    a. cardiovascular system _____

    b. digestive system _____

    c. respiratory system _____

    d. urinary system (i.e., the kidneys) _____

28. The two systems of the body that control homeostasis are a. _____

    and b. _____.

29. Figure 4.11 on pages 66–67 of your textbook illustrates the ways that the systems of the body contribute to homeostasis. Match the descriptions to the organ systems.

integumentary system   respiratory system   lymphatic system   digestive system
skeletal system   nervous system   urinary system   reproductive system
endocrine system   muscular system   cardiovascular system

a. _____ Breakdown and absorption of food materials.
b. _____ Gaseous exchange between external environment and blood.
c. _____ Regulation of all body activities; learning and memory.
d. _____ Body movement; production of body heat.
e. _____ External support and protection of body.
f. _____ Secretion of hormones for chemical regulation.
g. _____ Production of sperm; transfer of sperm to female reproductive system.
h. _____ Transport of nutrients to body cells; removal of wastes from cells.
i. _____ Immunity; absorption of fats; drainage of tissue fluid.
j. _____ Maintenance of volume and chemical composition of blood.
k. _____ Internal support and protection; body movement; production of blood cells.

DEFINITIONS WORDSEARCH

Review key terms by completing this wordsearch using the following alphabetized list of terms:

```
E H J Y R D E R M I S Y T X
P T W F E D P P O P U L I C
I W E D I L I G A M E N T L
T A S R T W D I O P T L N J
H M I B O N E G U B V E R O
E J T L B C R N O R U E N L
L C O E L O M K R L D K I D
I O F E T I I H U G N U F O
A M S E F Y S T R I A T E D
L P M I N D L E S H L I F B
C A R T I L A G E O G I F T
D C P T J U N E T F D I L T
C T E N D O N J U N P T I F
```

bone
cartilage
coelom
compact
dermis
epidermis
epithelial
gland
ligament
neuron
striated
tendon

a. _____ Type of tissue that covers the body and lines its cavities.

b. _____ Inner region of skin that lies beneath the epidermis.

c. _____ Fibrous connective tissue joining bone to bone.

d. _____ Rigid connective tissue containing mineral salts.

e. _____ Outer region of skin, protective in nature.

f. _____ Specialized epithelial cell that secretes substances.

g. _____ Having alternating light and dark bands.

h. _____ Connective tissue in which the cells lie in small chambers called lacunae, separated by matrix that is solid, yet flexible.

i. _____ Fibrous connective tissue connecting muscle to bone.

j. _____ Type of bone with densely packed osteons.

k. _____ Embryonic body cavity that becomes the thoracic and abdominal cavities.

l. _____ A nerve cell.

Do not refer to the text when taking this test.

_____1. Cells of a similar structure working together constitute a(n)
   a. tissue.
   b. organ.
   c. organism.
   d. organ system.

_____2. Epithelial tissues do which of the following?
   a. secrete
   b. line body cavities
   c. protect
   d. All of these are correct.

_____3. Which of these pairs is mismatched?
   a. fat–subcutaneous layer
   b. receptors–dermis
   c. keratinization–epidermis
   d. nerves and blood vessels–epidermis

_____4. If epithelial tissue is made up of many layers of cells, what is true?
   a. It is called simple.
   b. It is called stratified.
   c. It is called striated.
   d. It is pseudostriated.

_____5. Which type of epithelial tissue is composed of flattened cells?
   a. glandular
   b. squamous
   c. columnar
   d. cuboidal

_____6. What type of tissue supports epithelium?
   a. fibrous connective tissue
   b. nervous tissue
   c. muscular tissue
   d. loose connective tissue

_____7. Which type of cell junction allows cells to stretch and bend?
   a. tight junction
   b. gap junction
   c. adhesion junction
   d. intercalated disks

_____8. Which of the following tissues has cells residing in lacunae?
   a. adipose tissue
   b. fibrous connective tissue
   c. hyaline cartilage
   d. loose connective tissue

_____9. Osteocytes are residents of _____ tissue.
   a. cartilage
   b. bone
   c. muscle
   d. pseudostratified columnar epithelium

For questions 10–13, match the statements to these blood components:
   a. red blood cells
   b. white blood cells
   c. platelets

_____10. Contribute to blood clotting process
_____11. Carry oxygen to cells
_____12. Contain hemoglobin
_____13. Participate in immunity

In questions 14–17, match the statements to these components of skin:
   a. epidermis   b. arrector pili   c. melanocytes
   d. oil glands

_____14. When these become blocked, a blackhead forms.
_____15. Give rise to skin pigment
_____16. Muscle attached to a hair follicle
_____17. Keratinized layer of skin

_____18. Cells in the nervous tissue that support and protect neurons are called
   a. neurons.
   b. fibers.
   c. axons.
   d. neuroglia.

_____19. Cardiac muscle fibers are characterized by which traits?
   a. striated; intercalated disks
   b. smooth; single nucleus
   c. striated; multiple nuclei; intercalated disks
   d. smooth; tapered; multiple nuclei

_____20. Which of these is located in the thoracic cavity?
   a. small intestine
   b. urinary bladder
   c. lungs
   d. kidneys

_____21. The thoracic cavity is lined with
   a. mucous membrane.
   b. serous membrane.
   c. cutaneous membrane.
   d. synovial membrane.

_____22. In a negative feedback mechanism,
  a. homeostasis is impossible.
  b. there is a constancy of the internal environment.
  c. a wide fluctuation occurs continuously.
  d. None of the above is true.

_____23. When body temperature rises,
  a. sweat glands become active, and cool the body.
  b. sweat glands become inactive and cool the body.
  c. sweat glands become active and heat the body.
  d. None of the above is correct.

THOUGHT QUESTIONS

Answer in complete sentences.

24. How do the nervous system and endocrine system interact to maintain the homeostasis of the body?

25. How do the blood vessels, the digestive tract, the lungs, and the kidneys work together to maintain homeostasis?

Test Results: _____ number correct ÷ 25 = _____ × 100 = _____ %

ANSWER KEY

STUDY QUESTIONS

1. a. fibroblasts **b.** collagen **c.** elastin **2. a.** dense fibrous connective **b.** muscle **c.** bone **3. a.** Hyaline **b.** lacunae **c.** matrix **4. a.** elastic **b.** elastic **c.** Fibrocartilage **5.** See Figure 4.2, in text. **6. a.** Red **b.** White **c.** platelets **7. a.** striated; skeleton; voluntary **b.** spindle-shaped; internal organs; involuntary **c.** striated; heart; involuntary **8. a.** neurons **b.** nerves **c.** conduct nerve impulses **d.** neuroglia **e.** protection **f.** nine **g.** one-half **9. a.** shape **b.** Cuboidal **c.** columnar **10. a.** See Figure 4.8, in text. **b.** air sacs or blood vessels of the lungs **c.** absorption **d.** stratified **11.** This type of tissue appears to be layered but is not, and cells have small, hairlike projections called cilia. **12. a.** Gap **b.** Adhesion **c.** Tight **13. a.** epidermis **b.** keratin **14. a.** hair follicle **b.** Arrector pili **c.** Oil **15. a.** epidermis **b.** dermis **c.** subcutaneous layer **d.** hair shaft **e.** arrector pili muscle **f.** oil gland **g.** sensory receptors **h.** hair root **i.** adipose tissue **j.** blood vessels **k.** sweat gland **l.** nerve **16.** Sweat absorbs body heat as it evaporates. **17. a.** M **b.** C **c.** S **d.** R **e.** M **18. a.** digestive **b.** muscular **c.** nervous **d.** endocrine **19. a.** A **b.** D **c.** A **d.** T **e.** T **f.** A **g.** A **h.** A **20. a.** coelom **b.** thoracic **c.** abdominal **21. a.** meninges **b.** Mucous membrane **c.** Synovial **d.** synovial fluid **22. a.** serous **b.** watery serous fluid **23. a.** relative constancy of the internal environment. **b.** body temperature remains around 37°C. **24. a.** stimulus **b.** sensor **c.** control center **d.** effect **e.** homeostasis **25.** When the pancreas detects that the blood glucose level is too high, it secretes insulin, the hormone that causes cells to take up glucose. When the blood sugar level returns to normal, the pancreas is no longer stimulated to secrete insulin. **26. a.** F **b.** T **c.** T **d.** F **27. a.** refreshes tissue fluid **b.** provides nutrient molecules **c.** removes carbon dioxide and adds oxygen to the blood **d.** eliminates wastes and salts **28. a.** nervous **b.** endocrine **29. a.** digestive system **b.** respiratory system **c.** nervous system **d.** muscular system **e.** integumentary system **f.** endocrine system **g.** reproductive system **h.** cardiovascular system **i.** lymphatic system **j.** urinary system **k.** skeletal system

DEFINITIONS WORDSEARCH

```
E           D E R M I S
P           P
I           L I G A M E N T
T           D
H       B O N E
E           R N O R U E N
L C O E L O M       D
I     O         I       N
A     M     S T R I A T E D
L     P                 L
C A R T I L A G E   G
      C
      T E N D O N
```

a. epithelial   b. dermis   c. ligament   d. bone   e. epidermis   f. gland   g. striated   h. cartilage   i. tendon   j. compact   k. coelom   l. neuron

CHAPTER TEST

1. a   2. d   3. d   4. b   5. b   6. d   7. c   8. c   9. b   10. c   11. a   12. a   13. b   14. d   15. c   16. b   17. a   18. d   19. a   20. c   21. b   22. b   23. a   24. The nervous system is the ultimate source of control over homeostasis—it registers changes and takes appropriate actions. Nervous impulses can be sent to trigger an appropriate response to maintain homeostasis. The endocrine system releases hormones into the bloodstream when needed to maintain homeostasis. Hormones last longer, but nervous impulses can travel faster. The nervous system controls the endocrine system.   25. The digestive tract adds nutrients, and the lungs add oxygen to the blood. The blood vessels bring nutrients and oxygen to tissue fluid and take away waste molecules, including carbon dioxide. The kidneys excrete metabolic wastes, and the lungs excrete carbon dioxide.

5

CARDIOVASCULAR SYSTEM: HEART AND BLOOD VESSELS

STUDY TIPS

Learning about the anatomy and physiology of the heart is one of the first important steps one can take toward keeping healthy and fit. In this chapter, you will learn about the cardiovascular system, including the blood vessels, heart, pathways of circulation, and circulatory disorders.

When you study the pathway of circulation through the heart (p. 81), remember that blood flows first into the atria (A) and then into the ventricles (V). (Note that A comes before V in the alphabet.) The heart is slightly off center to the left side of the chest. This means the right side of the heart is closer to the midline and lies be-

tween the lungs. The right ventricle pumps blood to the lungs. O_2-rich blood returns to the left atrium. The left ventricle, with its thicker, muscular walls, pumps this blood to the remainder of the body.

Use Figure 5.9 to identify the causes of blood flow in the arteries as opposed to veins. The cardiovascular system contributes to homeostasis by servicing the cells. The pulmonary circuit makes blood O_2-rich, and the hepatic portal vein brings nutrients into the systematic circuit (pp. 86–87). Compare the structure of lymphatic vessels (p. 92) with that of the cardiovascular veins.

TIP: Be sure to visit the Online Learning Center that accompanies *Human Biology* 9/e. It has practice quizzes, interactive activities, labeling exercises, art quizzes, animations, flash cards, and much more. http://www.mhhe.com/maderhuman9

STUDY QUESTIONS

Study the text section by section. Answer the study questions so that you can fulfill the learning objectives for each section.

5.1 OVERVIEW OF THE CARDIOVASCULAR SYSTEM (PAGE 78)

After you have answered the questions for this section, you should be able to
• Name and describe the functions of the cardiovascular system.

1. The cardiovascular system consists of two components: the a._____, which pumps blood, and the the b._____, through which blood flows.

5.2 THE BLOOD VESSELS (PAGE 79)

After you have answered the questions for this section, you should be able to
• Name and describe the structure and function of arteries, capillaries, and veins.
• Explain how blood flow may bypass certain capillary beds.

2. Label the blood vessels in the following diagram using the alphabetized list of terms.

arterioles
artery
capillaries
heart
vein
venules

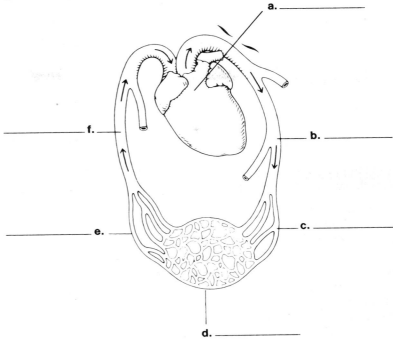

a. _____

f. _____

b. _____

e. _____

c. _____

d. _____

3. Match the statements to these terms.

artery   vein   capillary

a. _____   Has the thickest walls.
b. _____   Has valves.
c. _____   Takes blood away from the heart.
d. _____   Takes blood to the heart.
e. _____   Exchanges carbon dioxide and oxygen with tissues.
f. _____   Nervous stimulation causes these to constrict during hemorrhaging; they also act as a blood reservoir.

Use the space provided to answer this question in complete sentences.

4. Explain how it is possible to bypass capillary beds by shunting blood directly from arteriole to venule.

_____

_____

5.3 THE HEART (PAGES 80—83)

After you have answered the questions for this section, you should be able to
- Name the parts of the heart and their functions.
- Trace the path of blood flow through the heart.
- Explain how the conduction system of the heart controls the heartbeat.
- Describe how hormones and impulses from the nervous system and hormones can modify the heart rate.
- Label and explain a normal electrocardiogram.

5. a. Trace the path of blood through the heart from the vena cava to the lungs. _____

_____

b. Trace the path of blood from the lungs to the aorta. _____

_____

6. Label the parts of the heart in the following diagram using the alphabetized list of terms.

aorta
semilunar valves
atrioventricular
   (mitral) valve
atrioventricular
   (tricuspid) valve
chordae tendineae
inferior vena cava
left atrium
left pulmonary artery
left ventricle
pulmonary trunk
right atrium
right pulmonary arteries
right pulmonary veins
right ventricle
semilunar valves
septum
superior vena cava

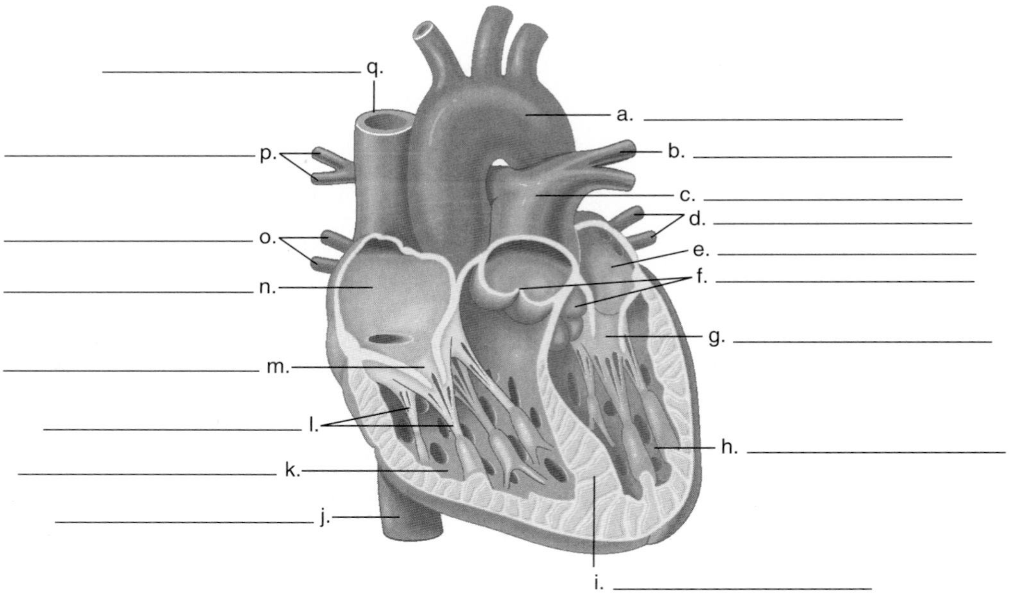

q. _____

a. _____

b. _____

c. _____

d. _____

e. _____

f. _____

p. _____

o. _____

n. _____

m. _____

l. _____

k. _____

j. _____

g. _____

h. _____

i. _____

Use the space provided to answer this question in complete sentences.

7. How does the thickness of the walls of the ventricles relate to their functions? _____
_____

8. The word $^{a.}$_____ refers to the working phase of the heart when the chambers contract, while
$^{b.}$_____ refers to the resting phase when the chambers are relaxed.

9. Heart sounds. When the atria contract, this forces blood through the $^{a.}$_____ valves into the
chambers called the $^{b.}$_____. The closing of these valves is the lub sound. Next, the ventricles
contract and force the blood into the arteries. Now the $^{c.}$_____ valves close, making the dub sound.

10. Match the phrases to these nodes:

    SA node   AV node

a. _____ pacemaker
b. _____ contraction of ventricles
c. _____ base of right atrium near the septum
d. _____ Purkinje fibers

11. Match the actions to these divisions of the nervous system:

    parasympathetic system   sympathetic system

a. _____ normal body functions
b. _____ active under times of stress
c. _____ releases norepinephrine to speed heart rate
d. _____ slows heart rate

12. Does the adrenal gland hormone epinephrine speed or slow the heart rate? _____

13. What is the significance of each of the following in an electrocardiogram?

    a. P wave _____

    b. QRS wave _____

    c. T wave _____

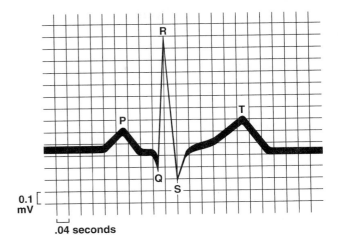

5.4 FEATURES OF THE CARDIOVASCULAR SYSTEM (PAGES 84–85)

After you have answered the questions for this section, you should be able to
- Identify the factors that influence blood pressure, pulse, and blood flow.

In questions 14–18, use the following diagram.

14. What force accounts for blood flow in arteries?

    _____

    _____

    _____

15. Why does this force fluctuate?

    _____

    _____

    _____

16. What causes the blood pressure and velocity to drop off?

    _____

    _____

    _____

17. What three factors account for blood flow in the veins?

    _____

    _____

    _____

18. What keeps blood from flowing backwards in veins?

    _____

5.5 The Cardiovascular Pathways (pages 86–87)

After you have answered the questions for this section, you should be able to

- Describe the pulmonary circuit of circulation.
- Identify the major vessels of the systemic circuit.

19. Trace the path of blood to the left atrium:

right ventricle

a. ______________________

b. ______________________

lungs

c. ______________________

left atrium

From the legs:

legs

d. ______________________

e. ______________________

right atrium

20. Trace the path of blood from the aorta to the liver:

aorta

a. ______________________

digestive tract

b. ______________________

liver

From the liver:

liver

c. ______________________

vena cava

5.6 Cardiovascular Disorders (pages 88–91)

After you have answered the questions for this section, you should be able to

- Discuss the factors that can lead to cardiovascular disease and how they can be prevented.

21. Match the statements to these items:

artificial pacemaker needed   dietary restriction of saturated fats and cholesterol
donor heart transplant   coronary bypass

a. ______________________ Clearing clogged arteries was unsuccessful.

b. ______________________ Blood pressure is 200/140.

c. ______________________ Heartbeat is irregular.

d. ______________________ Congestive heart failure is present.

22. Match the phrases to these items:

thrombus and embolus   atherosclerosis and hypertension   varicose veins   hemorrhoids

a. ______________________ stroke and heart attack

b. ______________________ varicose veins in rectum

c. ______________________ weakened valves of veins

d. ______________________ development of aneurysm

5.7 The Lymphatic System Helps the Cardiovascular System (page 92)

After you have answered the questions for this section, you should be able to

- List three functions of the lymphatic system.
- Describe the structure of lymphatic vessels.
- Explain the structure and purpose of lymph nodes.

23. Give three functions of the lymphatic system.

a. ______________________

b. ______________________

c. ______________________

24. Movement of lymph within _____ is dependent upon skeletal muscle contractions, respiratory movements, and valves.

25. Small ovoid or round structures that cleanse the lymph of debris and pathogens are called
    a. white blood cells.
    b. lymph nodes.
    c. the thoracic duct.
    d. the thymus.

DEFINITIONS WORDSEARCH

Review key terms by completing this wordsearch using the following alphabetized list of terms:

```
M Y O C A R D I U M L I M
O U I L T R G H D C V U N
Q R G T R G J K I N D E I
I C A P I L L A R Y R E E
R A T R O A L O P E R V V
D E C V V S G E T H E L Y
E Y S D E I Y X Z E R Y R
E R N J N H U S J L O P A
L E G U T A S D T A N D N
O T G B R M H K I O G O O
T R D E I N U N S U L L M
S A G Y C G A Y D O T E L
A Y S E U S A I D F U G U
I R R T L D V S K O P L P
D A D A A R T E R I O L E
M N A M R A W P I L L L A
U O R A V J I T M J E N N
I R F E A R M U A E D E A
N O E I L C H M A R D M A
E C M I V M I C O T T O N
A L C H E C H R W E N D Y
```

aorta
arteriole
atrioventricular valve
capillary
coronary artery
diastole
myocardium
pulmonary vein
septum
systole

a. _____ Valve located between the atrium and the ventricle.
b. _____ Vessel that takes blood from an artery to capillaries.
c. _____ Cardiac muscle in the wall of the heart.
d. _____ Partition in heart that divides it into left and right halves.
e. _____ Contraction of a heart chamber.
f. _____ Relaxation of a heart chamber.
g. _____ Blood vessel returning from lungs to heart.
h. _____ Large artery leaving heart with blood from the left ventricle.
i. _____ Microscopic vessel connecting arterioles to venules.
j. _____ Artery that supplies blood to the wall of the heart.

Do not refer to the text when taking this test.

____ 1. Which type of blood vessel allows exchange of material between the blood and the tissues?
   a. arteries
   b. arterioles
   c. capillaries
   d. veins
   e. venules

____ 2. What is the function of the heart valves?
   a. to push blood
   b. to prevent backflow
   c. to stimulate the heart
   d. to give support to the heart

____ 3. Arteries
   a. carry blood away from the heart.
   b. carry blood toward the heart.
   c. have valves.
   d. Both a and b are correct.

____ 4. Shunting of blood is possible because of a thoroughfare channel that directly joins _____ to _____.
   a. arterioles; venules
   b. venules; capillaries
   c. arteries; veins
   d. arterioles; veins

____ 5. Which of these vessels have thinner walls?
   a. arteries
   b. veins
   c. Both are the same.

____ 6. The venae cavae
   a. carry blood to the right atrium.
   b. carry blood away from the right atrium.
   c. join with the aorta.
   d. have a high blood pressure.

____ 7. The chamber of the heart that receives blood from the pulmonary veins
   a. is the right atrium.
   b. is the left atrium.
   c. contains O_2-rich blood.
   d. contains O_2-poor blood.
   e. a and c
   f. b and c
   g. b and d

____ 8. The coronary arteries carry blood
   a. from the aorta to the heart tissues.
   b. from the heart to the brain.
   c. directly to the heart from the pulmonary circuit.
   d. from the lungs directly to the left atrium.

____ 9. Which of these chambers has the thickest walls?
   a. right atrium
   b. right ventricle
   c. left atrium
   d. left ventricle

____ 10. When the atria are contracting, the ventricles are
   a. contracting.
   b. relaxing.
   c. in diastole.
   d. in systole.
   e. a and c
   f. b and c
   g. b and d

____ 11. The SA node
   a. works only when it receives a nerve impulse.
   b. is located in the left atrium.
   c. initiates the heartbeat.
   d. All of these are correct.

____ 12. The first wave (the P wave) of an ECG occurs prior to
   a. atrial contraction.
   b. ventricular contraction.
   c. ventricular relaxation.
   d. atrial relaxation.

____ 13. The heart sounds are due to
   a. blood flowing.
   b. the closing of the valves.
   c. the heart muscle contracting.
   d. blood pressure in the aorta.

____ 14. Blood flows in veins because of
   a. contraction of valves.
   b. arterial blood pressure.
   c. capillary blood pressure.
   d. skeletal muscle contraction.

____ 15. *Systole* refers to the contraction of the
   a. major arteries.
   b. SA node.
   c. atria and ventricles.
   d. major veins.

____ 16. Blood pressure falls off drastically in the capillaries because the capillaries
   a. contain valves.
   b. become veins.
   c. have a large cross-sectional area.
   d. All of these are correct.

____ 17. Which of these correctly traces the path of blood from the left ventricle to the head?
   a. left ventricle, subclavian artery, head
   b. left ventricle, pulmonary artery, head
   c. left ventricle, aorta, carotid artery, head
   d. left ventricle, vena cava, jugular vein, head

____ 18. Blood pressure
   a. is the same in all blood vessels.
   b. is highest in the aorta.
   c. is measured by taking an ECG.
   d. never rises above normal.

___ 19. The major portion of the cardiovascular system is called the _____ circuit.
   a. systemic
   b. pulmonary
   c. hepatic portal
   d. coronary

___ 20. Blood flowing to the lungs leaves the heart via the _____ and returns to the heart via the _____.
   a. aorta; superior vena cava
   b. superior vena cava; aorta
   c. pulmonary arteries; pulmonary veins
   d. aorta; pulmonary veins

___ 21. The jugular vein carries blood from the
   a. head.
   b. vena cava.
   c. arm.
   d. aorta.

___ 22. Blood moves slowly in capillaries,
   a. which facilitates tissue exchange.
   b. because they have valves.
   c. because they have thick walls.
   d. because venules are smaller than capillaries.

___ 23. The two collecting ducts of the lymphatic system empty into
   a. cardiovascular arteries.
   b. cardiovascular veins.
   c. pulmonary arteries.
   d. pulmonary veins.

___ 24. The structure of a lymphatic vessel is most similar to that of a
   a. cardiovascular artery.
   b. cardiovascular arteriole.
   c. cardiovascular vein.
   d. skeletal muscle fiber.

___ 25. Lymph is _____ in lymphatic vessels.
   a. blood
   b. serum
   c. tissue fluid
   d. plasma

___ 26. People with atherosclerosis often experience
   a. high blood pressure.
   b. a heart attack.
   c. a thrombus.
   d. a stroke.
   e. All of these are correct.

___ 27. A heart attack is due to a blocked
   a. pulmonary artery.
   b. coronary artery.
   c. aorta.
   d. vena cava.

___ 28. Phlebitis and hemorrhoids are conditions involving
   a. arteries.
   b. capillaries.
   c. veins.
   d. arterioles.

THOUGHT QUESTIONS

Use the space provided to answer these questions in complete sentences.

29. Why can arteries expand without rupturing?

30. Explain how the digestive system and cardiovascular system benefit each other.

Test Results: _____ number correct ÷ 30 = _____ × 100 = _____%

STUDY QUESTIONS

1. a. heart **b.** blood vessels **2. a.** heart **b.** artery **c.** arterioles **d.** capillaries **e.** venules **f.** vein **3. a.** artery **b.** vein **c.** artery **d.** vein **e.** capillary **f.** vein **4.** The shunting of blood around capillary beds is possible because each bed has a thoroughfare channel that allows blood to flow directly from arteriole to venule. Sphincter muscles prevent blood from flowing into the capillaries. **5. a.** vena cava, right atrium, atrioventricular valve, right ventricle, pulmonary semilunar valve, pulmonary artery, lungs **b.** lungs, pulmonary veins, left atrium, atrioventricular valve, left ventricle, aortic semilunar valve, aorta **6. a.** aorta **b.** left pulmonary artery **c.** pulmonary trunk **d.** left pulmonary veins **e.** left atrium **f.** semilunar valves **g.** atrioventricular (mitral) valve **h.** left ventricle **i.** septum **j.** inferior vena cava **k.** right ventricle **l.** chordae tendineae **m.** atrioventricular (tricuspid) valve **n.** right atrium **o.** right pulmonary veins **p.** right pulmonary arteries **q.** superior vena cava; see also Fig. 5.4, p. 81, in text. **7.** The left ventricle is thicker-walled than the right ventricle because the left one must pump blood the greater distance to the entire body. The right ventricle only pumps the shorter distance to the lungs. **8. a.** *systole* **b.** *diastole* **9. a.** atrioventricular **b.** ventricles **c.** semilunar **10. a.** SA node **b.** AV node **c.** AV node **d.** AV node **11. a.** parasympathetic system **b.** sympathetic system **c.** sympathetic system **d.** parasympathetic system **12.** speed **13. a.** associated with atrial systole **b.** associated with ventricular systole **c.** associated with ventricular recovery **14.** Blood pressure created by contraction of the heart accounts for blood flow in arteries. **15.** The heart contracts during systole and rests during diastole. When the heart rests, blood pressure drops off. **16.** Distance from heart and an increase in total cross-sectional area cause blood pressure and velocity to drop off. **17.** Skeletal muscle contraction, respiratory movements, and presence of valves account for blood flow in the veins. **18.** Valves keep the blood from flowing backwards in the veins. **19. a.** pulmonary trunk **b.** pulmonary arteries **c.** pulmonary veins **d.** iliac vein **e.** inferior vena cava **20. a.** mesenteric arteries **b.** hepatic portal vein **c.** hepatic vein **21. a.** coronary bypass **b.** dietary restriction **c.** artificial pacemaker needed **d.** donor heart transplant **22. a.** thrombus and embolus; atherosclerosis and hypertension **b.** hemorrhoids **c.** varicose veins **d.** atherosclerosis and hypertension **23. a.** return of excess tissue fluid to bloodstream **b.** receive lipoproteins at intestinal villi **c.** defense against disease **24.** lymphatic vessels **25.** b

DEFINITIONS WORDSEARCH

```
        M Y O C A R D I U M
                T               N
                R               I
        C A P I L L A R Y       E
        A T R O A               V
          V S                   Y
        Y E Y                   R
      E R N         S           A
      L E T         T           N
      O T R             O       O
      T R I             L       M
      S A C               E     L
      A Y U                     U
      I R L         S           P
      D A A R T E R I O L E
        N R         P
        O V         T
        R A         U
        O L         M
        C V
          E
```

a. atrioventricular valve **b.** arteriole **c.** myocardium **d.** septum **e.** systole **f.** diastole **g.** pulmonary vein **h.** aorta **i.** capillary **j.** coronary artery

CHAPTER TEST

1. c **2.** b **3.** a **4.** a **5.** b **6.** a **7.** f **8.** a **9.** d **10.** f **11.** c **12.** a **13.** b **14.** d **15.** c **16.** c **17.** c **18.** b **19.** a **20.** c **21.** a **22.** a **23.** b **24.** c **25.** c **26.** e **27.** b **28.** c **29.** The wall of an artery has a middle layer of elastic tissue and smooth muscle and an outer layer of fibrous connective tissue. This arrangement allows for both strength and flexibility. **30.** The cardiovascular system transports nutrients from the digestive tract to the rest of the body and services the organs of the digestive tract. The digestive tract provides nutrients for blood cell formation and the formation of plasma proteins. The liver detoxifies the blood, makes plasma proteins, and destroys old red blood cells.

6

CARDIOVASCULAR SYSTEM: BLOOD

STUDY TIPS

Plasma (p. 99) carries nutrients as well as the proteins and platelets for producing blood clotting in response to injury and blood loss. Red blood cells contain the pigment hemoglobin (p. 100) that distributes oxygen throughout the body. White blood cells (p. 103) are responsible for fighting pathogens. You will learn more about the functions of white blood cells in Chapter 21. After reading this chapter, you will more fully appreciate why you need to eat adequately to maintain a healthy internal environment and, therefore, homeostasis. The critical functions performed for you by your blood help you survive.

When you study the mechanism of blood clotting (p. 104), make a flow diagram to help you remember what factor activates the next one (see Fig. 6.5 in the text). Draw and carefully label another diagram showing how exchange of materials occurs between the blood and tissue fluid (see Fig. 6.9 in the text). Toward the chapter's end, you will learn about the ABO system and blood typing (p. 106). This will help you to understand why it is important for medical personnel to be careful in administering transfusions.

TIP: Be sure to visit the Online Learning Center that accompanies *Human Biology* 9/e. It has practice quizzes, interactive activities, labeling exercises, art quizzes, animation, flash cards, and much more. http://www.mhhe.com/maderhuman9

STUDY QUESTIONS

Study the text section by section. Answer the study questions so that you can fulfill the learning objectives for each section.

6.1 BLOOD: AN OVERVIEW (PAGE 99)

After you have answered the question for this section, you should be able to
• Describe the composition and general functions of blood.

1. Place the appropriate letter pertaining to a function of the blood next to each statement.
   T—transport
   D—defense
   R—regulation

   a. _____ Certain blood cells are capable of engulfing and destroying pathogens.
   b. _____ Blood picks up carbon dioxide and wastes at the tissues.
   c. _____ The salts and plasma proteins help maintain its salt/water balance.

6.2 COMPOSITION OF BLOOD (PAGES 99–103)

After you have answered the questions for this section, you should be able to
• Describe the composition of plasma, and give several functions for plasma proteins.
• List the characteristics of red blood cells.
• Understand how the structure of hemoglobin allows it to carry oxygen throughout the body.
• Describe the life cycle of red blood cells.
• List the types of anemia and their causes.
• List the various types of white blood cells, and describe their structures and unique functions.

2. Plasma is mostly a._____, b._____ and c._____.

3. Place the correct plasma protein in the blank: Fibrinogen, Albumin, Globulin, or All plasma proteins.

   a._____ transport(s) cholesterol.

   b._____ help(s) blood clot.

   c._____ transport(s) bilirubin.

   d._____ help(s) maintain the pH and osmotic pressure of the blood.

4. Circle the items that correctly describe hemoglobin:
   a. heme contains iron
   b. globin contains iron
   c. becomes oxyhemoglobin in the tissues
   d. becomes deoxyhemoglobin in the tissues
   e. makes red blood cells red
   f. makes eosinophils red

5. The red blood cells, scientifically called ᵃ·_____, are made in the ᵇ·_____. Upon maturation, they are biconcave disks that lack a(n) ᶜ·_____ and contain ᵈ·_____. After about 120 days, red blood cells are destroyed in the ᵉ·_____ and _____. The condition of ᶠ·_____ is characterized by an insufficient number of red blood cells or not enough hemoglobin.

6. What molecule stimulates the production of red blood cells? _____

7. White blood cells, scientifically called ᵃ·_____, are made in the ᵇ·_____.

8. Name three differences between red blood cells and white blood cells. White blood cells are ᵃ·_____ in size than red blood cells; they do have a ᵇ·_____; and they do not contain ᶜ·_____.

9. Place the name of the correct white blood cell (neutrophil, eosinophil, basophil, lymphocyte, or monocyte) next to the proper description.

   a. _____ An agranular cell with a large, round nucleus that occurs in two versions. The B cells produce antibodies and the T cells destroy cells that contain viruses.

   b. _____ An abundant granular cell with a multilobed nucleus that phagocytizes pathogens.

   c. _____ A large, agranular cell that takes up residence in the tissues and differentiates into a voracious macrophage.

   d. _____ A cell with blue-staining granules that occurs in connective tissue and, like mast cells, releases histamine.

   e. _____ A cell with a bilobed nucleus and red-staining granules that becomes abundant during allergies and parasitic infections.

10. Platelets are
    a. involved in clotting.
    b. also called thrombocytes.
    c. derived from megakaryocytes.
    d. All of the above are correct.

6.3 BLOOD CLOTTING (PAGE 104)

After you have answered the questions for this section, you should be able to
• Explain how clotting factors in the blood operate to produce a blood clot.

11. The following shows the reactions that create blood clots:

    platelets ————————→ form plug
    prothrombin ————————→ thrombin
    fibrinogen ————————→ fibrin threads

Does the left-hand side or the right-hand side list substances that are always present in the blood? a._____
Which substance functions as an enzyme? b._____ Which substance is the actual clot?
c._____

12. Several nutrients are necessary for clotting to occur. Vitamin a._____ is needed for the production of prothrombin. The element b._____ is needed for the conversion of prothrombin to thrombin.

6.4 BLOOD TYPING (PAGES 106–107)

After you have answered the questions for this section, you should be able to
- Describe how antibodies and antigens determine blood type in the ABO system.
- Describe the possible Rh factor complications of pregnancy.

13. The following diagram shows the results of typing someone's blood. What is the blood type? _____

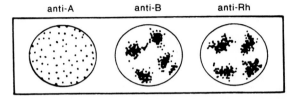

14. Draw a similar diagram showing the results if someone has AB-negative blood.

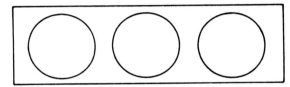

15. The following table indicates the blood types. Fill in the fourth and fifth columns by using this formula: The donor's antigen(s) must not be of the same type of letter as the recipient's antibody (antibodies).

Blood Type	Antigen	Antibody	Can Receive From	Can Donate To
A	A	Anti-B	a.	b.
B	B	Anti-A	c.	d.
AB	A, B	_____	e.	f.
O	None	Anti-A and B	g.	h.

Use the space provided to answer question 16 in complete sentences.

16. Consider these possible combinations of mates:
Rh+ mother and Rh− father   Rh− mother and Rh− father
Rh+ mother and Rh+ father   Rh− mother and Rh+ father

Which of these combinations can cause pregnancy difficulties, and why?

_____

_____

After you have answered the questions for this section, you should be able to
- Explain how the exchange of materials occurs between blood and tissue fluid.
- Discuss how exchange of substances between blood and tissue fluid across capillary walls supplies cells with nutrients and removes wastes.

17. Using Figure 6.9 as a guide, label the following diagram using these terms:

amino acids	glucose	oxygen
arterial end	net pressure in	tissue fluid
blood pressure (two	net pressure out	venous end
times)	osmotic pressure (two	wastes
carbon dioxide	times)	water (two times)

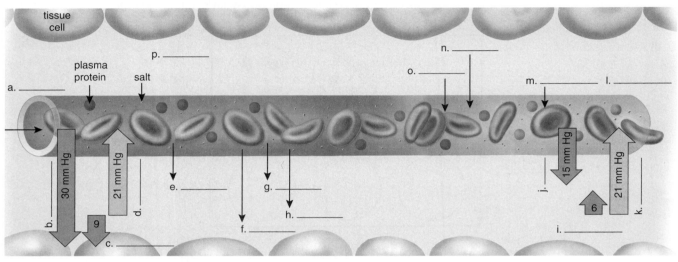

Use the space provided to answer questions 18–19 in complete sentences.

18. Use the diagram to explain capillary exchange. _____

_____

19. Why is there excess tissue fluid, and what happens to it?_____

_____

20. Label the following diagram, using the alphabetized list of terms.
    - blood capillary
    - blood flow
    - lymph flow
    - lymphatic capillary
    - tissue fluid flow

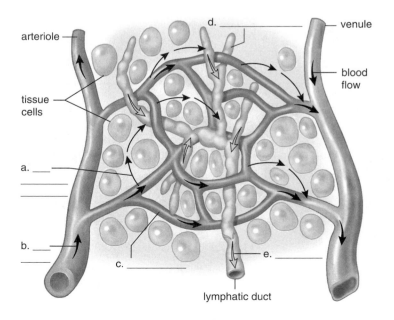

After you have answered the questions for this section, you should be able to
• Discuss how the cardiovascular system works with other systems of the body to maintain homeostasis.

21. The *Human Systems Work Together* diagram in your textbook indicates ways in which the cardiovascular system benefits other organ systems of the body. Match the organ systems with the correct descriptions.

(1) Lymphatic vessels collect excess tissue fluid and return it to blood vessels; lymphatic organs store lymphocytes; lymph nodes filter lymph and the spleen filters blood.
(2) Blood vessels deliver nutrients and oxygen to bones, carry away wastes.
(3) Blood vessels deliver nutrients and oxygen to muscles, carry away wastes.
(4) Blood vessels deliver wastes to be excreted; blood pressure aids kidney function; blood services urinary organs.
(5) Blood vessels deliver nutrients and oxygen to skin, carry away wastes; blood clots if skin is broken.
(6) Blood vessels transport hormones from glands; blood services glands; heart produces atrial natriuretic hormone.
(7) Blood vessels transport nutrients from digestive tract to body; blood services digestive organs.
(8) Blood vessels transport sex hormones; vasodilation causes genitals to become erect; blood services reproductive organs.
(9) Blood vessels deliver nutrients and oxygen to neurons, carry away wastes.
(10) Blood vessels transport gases to and from lungs; blood services respiratory organs.

a. _____ integumentary system
b. _____ respiratory system
c. _____ lymphatic system
d. _____ muscular system
e. _____ skeletal system
f. _____ nervous system
g. _____ urinary system
h. _____ reproductive system
i. _____ endocrine system
j. _____ digestive system

DEFINITIONS WORDSEARCH

Review key terms by completing this wordsearch using the following alphabetized list of terms:

```
A L R P M T Y H U J F V B U M P
A N E M I A F T G D C U M O L E
R T D O P L E T G I K I D T B L
A E B O L I K A T B G U J I Y L
G I L U K I N G T F C V R M I K
C L O T T I N G Y H U R P O L I
D E O R R T I L O O P H U N I A
R E D S D R U U E R T E R Y T M
H A C X W A Z T H J N M U R E S
F T E F T U K I P L O O M L L A
A L L D F E S N Y U G G H U E L
C O L D S E T A F V C L F Y T P
T R I P D Y H T F G I O O D A Z
O R G T U N T I J I O B G Y L F
R T R L I H P O S A B I A D P Y
T H R O M B I N I G H N D O P W
```

agglutination
anemia
basophil
clotting
hemoglobin
lymph
plasma
platelet
red blood cell
serum
thrombin

a. _____ Fluid derived from tissue fluid that circulates in lymphatic vessels.

b. _____ Light yellow liquid left after clotting of blood.

c. _____ Condition in which shortage of hemoglobin or red blood cells leads to oxygen shortage in blood.

d. _____ Granular white blood cell that can be stained with a basic dye.

e. _____ Liquid portion of blood.

f. _____ Enzyme that triggers the conversion of fibrinogen to fibrin during blood clotting.

g. _____ Erythrocyte.

h. _____ Clumping of red blood cells during antigen-antibody reactions.

i. _____ Iron-containing protein in red blood cells that carries oxygen.

j. _____ Cell fragment in blood that functions in clotting.

k. _____ Process of blood coagulation, usually when injury occurs.

CHAPTER TEST

OBJECTIVE TEST

Do not refer to the text when taking this test.

____ 1. Which of these is mismatched?
   a. red blood cells—transport bilirubin
   b. white blood cells—fight infection
   c. platelets—clotting
   d. plasma—transport nutrients

____ 2. Hemoglobin
   a. transports O_2.
   b. transports CO_2.
   c. is found in erythrocytes.
   d. All of these are correct.

____ 3. Which of the following characterizes anemia?
   a. low red blood cell count, low hemoglobin, or both
   b. viral infection
   c. congenital disease
   d. All of these are correct.

____ 4. When blood travels through capillary beds and gives off oxygen to tissue cells, hemoglobin can now be called
   a. oxyhemoglobin.
   b. deoxyhemoglobin.
   c. carboxyhemoglobin.
   d. deaminated.

____ 5. Aged red blood cells are destroyed in the
   a. red bone marrow.
   b. lungs.
   c. lymph nodes.
   d. spleen and liver.

____ 6. Which of the following white blood cells has granules and is phagocytic?
   a. lymphocyte
   b. basophil
   c. monocyte
   d. neutrophil

____ 7. Which of these is NOT a valid contrast between red blood cells and white blood cells?
   red   white
   a. erythrocyte–leukocyte
   b. phagocytic–motile
   c. lacks nucleus–has nucleus
   d. numerous–less numerous

____ 8. Choose the best description of neutrophils.
   a. multilobed nuclei, phagocytic, granules do not take up stain
   b. U-shaped nucleus, dark blue after staining, turn into mast cells
   c. bilobed nucleus, red after staining, present with allergies
   d. kidney-shaped nucleus, phagocytic, turn into macrophages

____ 9. Choose the best description of basophils.
   a. lobed nuclei, phagocytic, do not take up stain
   b. U-shaped nucleus, dark blue after staining, related to mast cells
   c. bilobed nucleus, red after staining, present with allergies
   d. kidney-shaped nucleus, phagocytic, turn into macrophages

____ 10. Plasma transports
   a. nutrients.
   b. CO_2.
   c. hormones.
   d. All of these are correct.

____ 11. Which plasma protein becomes the threads of a clot?
   a. prothrombin
   b. thrombin
   c. prothrombin activator
   d. fibrinogen

____12. Which of these is NOT a normal function of plasma proteins?
   a. maintaining osmotic pressure
   b. a widely used source of nutrition for the body
   c. fighting infection
   d. contributing to blood clotting

____13. Plasma is the
   a. same as tissue fluid.
   b. liquid remaining after blood clots.
   c. liquid portion of the blood.
   d. All of these are correct.

____14. At a capillary
   a. oxygen is exchanged for carbon dioxide.
   b. glucose is exchanged for amino acids.
   c. water is exchanged for proteins.
   d. waste material carried in blood is deposited into tissues.

____15. At a capillary
   a. glucose and oxygen exit from the midsection, while carbon dioxide and wastes enter at the midsection.
   b. glucose exits from the arterial end, and oxygen and carbon dioxide enter at the venous end.
   c. blood pressure increases as the cross-sectional area increases.
   d. glucose and oxygen exit from the arterial end, and carbon dioxide enters at the venous end.

____16. Water leaves capillaries at their arterial ends because
   a. osmotic pressure gradients are in opposite directions.
   b. blood pressure is greater than the osmotic pressure.
   c. a gradient is established for diffusion.
   d. osmotic pressure is always greater than blood pressure.

____17. Water reenters capillaries at their venous ends because of
   a. active transport from interstitial fluid.
   b. osmotic pressure of blood drawing fluid from tissues.
   c. increasing blood pressure.
   d. increasing hemoglobin production.

____18. Which mineral is needed for clotting?
   a. iron
   b. calcium
   c. thrombin
   d. manganese

____19. Which of these is needed for clotting?
   a. platelets
   b. vitamin K
   c. fibrinogen and prothrombin proteins
   d. All of these are correct.

____20. The last step in blood clotting
   a. requires potassium ions.
   b. occurs outside the bloodstream.
   c. converts thrombinogen to thrombin.
   d. converts fibrinogen to fibrin.

____21. Blood types are determined by the individual's
   a. antibodies.
   b. antigens.
   c. lymphocytes.
   d. phagocytes.

____22. A person with blood type O lacks
   a. antigens on the red blood cells.
   b. antigens in the plasma.
   c. the Rh factor.
   d. Both b and c are correct.

____23. The agglutination of red blood cells occurs whenever
   a. antibodies bind with antigens on red blood cells.
   b. a person receives a blood transfusion from someone with an incompatible blood type.
   c. complementary antibodies combine.
   d. blood cells are destroyed by leukocytes.
   e. Both a and b are correct.

____24. In which case will hemolytic disease of the newborn occur?
   a. Rh⁻ mother; Rh⁺ fetus
   b. Rh⁺ mother; Rh⁻ fetus
   c. A mother; O fetus
   d. Both a and b are correct.

____25. An Rh⁺ fetus being carried by an Rh⁻ mother
   a. develops antibodies to the mother's blood.
   b. develops antigens to the mother's blood.
   c. may have its red blood cells attacked by antibodies made by the mother.
   d. may have its red blood cells attacked by antigens made by the mother.

THOUGHT QUESTIONS

Use the space provided to answer these questions in complete sentences.

26. Who should not give blood?

27. What data regarding a person's plasma and red blood cells are needed before deciding who can donate blood to whom?

Test Results: _____ number correct ÷ 27 = _____ × 100 = _____%

ANSWER KEY

STUDY QUESTIONS

1. a. D **b.** T **c.** R **2. a.** water **b.** plasma proteins **c.** salts **3. a.** Globulin **b.** Fibrinogen **c.** Albumin **d.** All plasma proteins **4.** a, d, e **5. a.** erythrocytes **b.** red bone marrow **c.** nucleus **d.** hemoglobin **e.** liver, spleen **f.** anemia **6.** erythropoietin **7. a.** leukocytes **b.** red bone marrow **8. a.** larger **b.** nucleus **c.** hemoglobin **9. a.** lymphocyte **b.** neutrophil **c.** monocyte **d.** basophil **e.** eosinophil **10.** d **11. a.** left-hand side **b.** thrombin **c.** fibrin threads **12. a.** K **b.** calcium **13.** B⁺ **14.** clumping should occur for anti-A and anti-B. **15. a.** A, O **b.** A, AB **c.** B, O **d.** B, AB **e.** A, B, AB, O **f.** AB **g.** O **h.** A, B, AB, O **16.** Rh⁻ mother and Rh⁺ father, because the mother might form antibodies to destroy red blood cells of this or a future baby who is Rh⁺. **17. a.** arterial end **b.** blood pressure **c.** net pressure out **d.** osmotic pressure **e.** water **f.** oxygen **g.** amino acids **h.** glucose **i.** net pressure in **j.** blood pressure **k.** osmotic pressure **l.** venous end **m.** water **n.** wastes **o.** carbon dioxide **p.** tissue fluid. **18.** At the arterial end of a capillary, blood pressure is higher than osmotic pressure; therefore, water leaves a capillary. In the midsection, molecules follow their diffusion gradients; nutrients and O₂ exit, while CO₂ and wastes enter the capillary. At the venous end of a capillary, osmotic pressure is higher than blood pressure; therefore, water enters a capillary. **19.** This system never retrieves all the water that leaves capillaries, and excess tissue fluid is picked up by lymphatic vessels and returned to the bloodstream. **20. a.** tissue fluid flow **b.** blood flow **c.** blood capillary **d.** lymphatic capillary **e.** lymph flow **21. a.** 5 **b.** 10 **c.** 1 **d.** 3 **e.** 2 **f.** 9 **g.** 4 **h.** 8 **i.** 6 **j.** 7

DEFINITIONS WORDSEARCH

```
            R
  A N E M I A                          L
            D              A           Y
            B              A           Y
            L              G       M
  C L O T T I N G              P
            O              L       H       A
            D              U       E       T   M
            C              T       M U R E S
            E              I       O       L   A
            L              N       G       E   L
            L              A       L       T   P
                           T       O       A
                           I       B       L
                 L I H P O S A B I         P
  T H R O M B I N                  N
```

a. lymph **b.** serum **c.** anemia **d.** basophil **e.** plasma **f.** thrombin **g.** red blood cell **h.** agglutination **i.** hemoglobin **j.** platelet **k.** clotting

CHAPTER TEST

1. a **2.** d **3.** a **4.** b **5.** d **6.** d **7.** b **8.** a **9.** b **10.** d **11.** d **12.** b **13.** c **14.** a **15.** a **16.** b **17.** b **18.** b **19.** d **20.** d **21.** b **22.** a **23.** e **24.** a **25.** c **26.** Individuals who have a disease that would harm a recipient should not give blood. **27.** To decide who can give blood to whom, you need to know the antigens on the red blood cells of the potential donors and the antibodies in the plasma of the potential recipients.

7

DIGESTIVE SYSTEM AND NUTRITION

Begin your study of this chapter by surveying the organs of the digestive system (pp. 116–123), and discover how each is uniquely adapted to carrying out a function that contributes to the overall process of digestion. Next, study the contributions of accessory organs (pp. 124–125) and digestive enzymes (pp. 126–127) that complete the chemical breakdown of nutrients.

A knowledge of nutrition is essential for good health and disease prevention. More information comes to light daily about what we should and should not eat.

Following the basic nutrition guidelines set forth in this chapter (pp. 128–137) will contribute to a healthy lifestyle.

Keep in mind as you study this chapter that the digestive system contributes to homeostasis by breaking down complex foods into nutrients that can be used by your body's *cells*.

TIP: Be sure to visit the Online Learning Center that accompanies *Human Biology* 9/e. It has practice quizzes, interactive activities, labeling exercises, art quizzes, animation, flash cards, and much more. http://www.mhhe.com/maderhuman9

STUDY QUESTIONS

Study the text section by section. Answer the study questions so that you can fulfill the learning objectives for each section.

7.1 THE DIGESTIVE TRACT (PAGES 116–123)

After you have answered the questions for this section, you should be able to
- Trace the path of food from mouth to anus during digestion.
- Describe the features of the mouth that prepare food for swallowing.
- Explain how the structure of the pharynx ensures the passage of food into the esophagus.
- Discuss the features of the esophagus and how peristalsis moves food along this structure.
- Describe the structure and function of the four layers of the digestive tract.
- Explain how the structure of the stomach is adapted to high acidity.
- Relate the function of the small intestine to its unique structure.
- Explain how villi and microvilli enhance absorption.
- Discuss how digestive secretions are under hormonal control, and list the hormones involved.
- Describe the sections of the large intestine, and know its general function.

Study the following diagram to refresh your memory of the path of food and the digestive process.

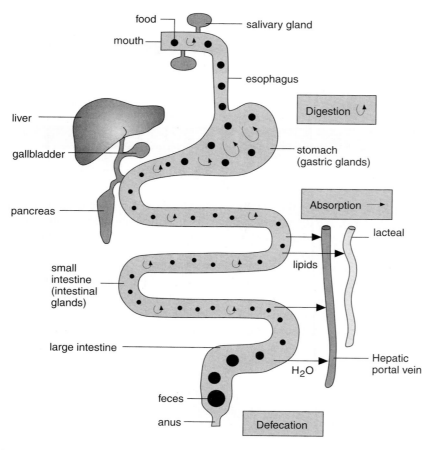

In questions 1–7, fill in the blanks.

1. Food is received by the ᵃ·_____, which possesses ᵇ·_____ to grind and break up particles. ᶜ·_____ glands secrete ᵈ·_____, which moistens food and binds it together for swallowing. The taste of food is detected by ᵉ·_____ located on the ᶠ·_____.

2. At the back of the mouth, the muscular ᵃ·_____ is involved in swallowing food. The ᵇ·_____ folds over the glottis during swallowing.

3. The esophagus propels food toward the ᵃ·_____, using muscular contractions called ᵇ·_____. At the upper end of the esophagus, a muscular ring called a(n) ᶜ·_____ regulates the entrance of food to the esophagus.

4. The layer of the digestive tract that houses blood vessels is the ᵃ·_____. The ᵇ·_____ layer contains two layers of smooth muscle. The outside layer, ᶜ·_____, secretes a fluid that enables organs to slide against one another.

5. The stomach is lined with ᵃ·_____, which secrete protective ᵇ·_____. Otherwise, the lining would be injured by the strong ᶜ·_____ acid secreted by ᵈ·_____ glands. Gastric juice also contains ᵉ·_____, which breaks down proteins.

6. The first section of the small intestine is the ᵃ·_____. It receives secretions from the ᵇ·_____ and ᶜ·_____ and receives food, known as ᵈ·_____, from the stomach.

7. The mucosa of the small intestine is folded into ᵃ·_____, which, in turn, have projections from individual cells called ᵇ·_____. ᶜ·_____ and ᵈ·_____ are absorbed directly into the bloodstream, and fats must be reconstructed so they can travel into vessels of the lymphatic system called ᵉ·_____.

Use the space provided to answer question 8 in a complete sentence.

8. What is the purpose of microvilli? _____

_____

In questions 9–10, fill in the blanks.

9. A number of hormones control the secretions of digestive juices. a. _____ is secreted in response to protein in foods and enhances gastric gland output, while b. _____ inhibits gastric secretion. c. _____ is secreted in response to acidic chyme.

10. The large intestine functions to store and compact a. _____. Unusual outgrowths of the lining of the colon, called b. _____, can be either benign or cancerous. Colon cancer incidence may increase for people who do not have enough c. _____ in their diets.

7.2 THREE ACCESSORY ORGANS (PAGES 124–125)

After you have answered the questions for this section, you should be able to
- Explain how the secretions from the pancreas aid digestion.
- List several functions of the liver, and describe the role of the gallbladder in digestion.
- Discuss the major diseases of the liver and their causes.

11. Pancreatic juice contains a mix of a. _____ solution to neutralize stomach acid and digestive b. _____ to further break down food.

12. The liver produces a greenish substance called a. _____, stored and concentrated by the b. _____. The liver has been called the c. _____ to the blood because it detoxifies substances entering the blood from the d. _____. Other functions of the liver include (list three)

e. _____

f. _____

g. _____

13. a. _____ is an inflammatory disease of the liver caused by a viral infection, while b. _____ is damage caused by chronic alcohol abuse.

7.3 DIGESTIVE ENZYMES (PAGES 126–127)

After you have answered the questions for this section, you should be able to
- Name the major digestive enzymes and the types of nutrients they digest, and state in which organs they are produced.

14. Complete the following table.

Major Digestive Enzymes

Food	Digestion	Enzyme	Optimum pH	Produced By	Site of Action
Starch	Starch + H_2O ⟶ maltose	a. _____ Pancreatic amylase	Neutral c. _____	Salivary glands d. _____	b. _____ Small intestine
	Maltose + H_2O ⟶ glucose + glucose	e. _____	Basic	Small intestine	f. _____
Protein	Protein + H_2O ⟶ peptides	Pepsin	g. _____	h. _____	Stomach
	Peptide + H_2O ⟶ amino acids	i. _____ Peptidases	Basic k. _____	Pancreas l. _____	j. _____ Small intestine
Nucleic acid	RNA and DNA + H_2O ⟶ nucleotides	m. _____	Basic	Pancreas	n. _____
	Nucleotide + H_2O ⟶ base + sugar + phosphate	Nucleosidases	o. _____	p. _____	Small intestine
Fat	Fat droplet + H_2O ⟶ glycerol + fatty acids	q. _____	Basic	Pancreas	r. _____

15. Match the enzyme to the food or breakdown product.

salivary amylase/pancreatic amylase   pepsin/trypsin   lipase   nuclease   maltase
peptidase   nucleosidases

a. _____ protein

b. _____ maltose

c. _____ nucleic acid

d. _____ starch

e. _____ fat

f. _____ nucleotides

g. _____ peptides

7.4 NUTRITION (PAGES 128–137)

After you have answered the questions for this section, you should be able to
- Describe how the different classes of nutrients enter into general circulation within the body.
- Discuss proper nutrition and how carbohydrates, protein, and lipids should be proportioned in the diet.
- Discuss the vitamin and mineral requirements in the diet.
- Describe the basis of several common eating disorders.

16. Fill in the blanks of the food guide pyramid in the following diagram recommended by nutritionists at the Harvard Medical School. Include the following groups, and indicate the recommended number of daily servings from each group:

     dairy or calcium supplement
     fish, poultry, and eggs
     fruit
     nuts and legumes
     plant oils
     red meat and butter
     vegetables
     white rice, white bread, potatoes, pasta, and sweets
     whole-grain foods

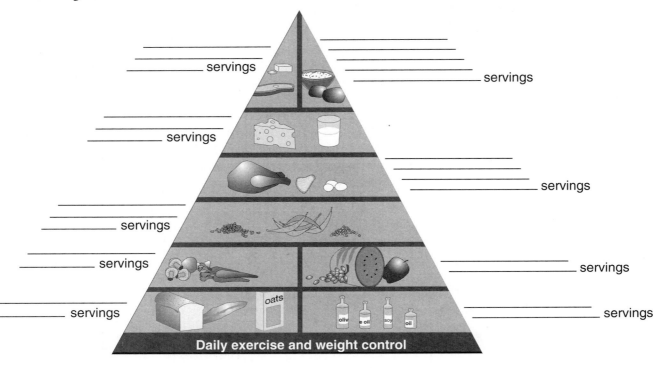

17. Place a check beside the most nutritional food.

    a. _____ fat

    b. _____ beef

    c. _____ vegetables

18. Place a check beside the type of food that should be the bulk of any diet.

    a. _____ vegetables

    b. _____ fruits

    c. _____ dairy

    d. _____ complex carbohydrates

19. Indicate whether these statements are true (T) or false (F).

    a. _____ Complex carbohydrates provide energy and fiber.

    b. _____ Meat provides proteins, but it also contains saturated fats.

    c. _____ The body can make all the essential amino acids it needs.

    d. _____ Saturated fats, whether in butter or margarine, can lead to more low-density lipoprotein (LDL), and this can cause plaque buildup in the arteries.

20. Many vitamins function as a. _____ in various metabolic pathways. Vitamin b. _____ is important as a precursor for a visual pigment, and vitamin c. _____ becomes a compound that enhances calcium absorption. Vitamins d. _____ are antioxidants.

21. Too much of the mineral a. _____ can lead to hypertension. The mineral b. _____ is a major component of bones and teeth.

22. Place the appropriate letter next to each statement.

    O—obesity   B—bulimia nervosa   A—anorexia nervosa

    a. _____ Body weight is normal, but weight is regulated by purging after eating.
    b. _____ Body weight is too low, and person may binge and purge.
    c. _____ Body weight is 20% or more above appropriate weight for height.
    d. _____ Exercise is usually minimal.
    e. _____ It is often accompanied by excessive exercise.

DEFINITIONS CROSSWORD

Review key terms by completing this crossword puzzle using the following alphabetized list of terms:

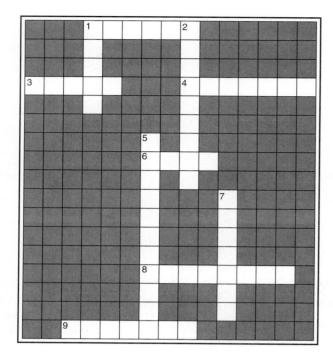

anus
duodenum
esophagus
fiber
gallbladder
gastric
lacteal
lipase
liver
pharynx

Across
1. Fat-digesting enzyme secreted by the pancreas.
3. Indigestible plant material that may lower risk of colon cancer.
4. Muscular passageway at the back of the mouth where swallowing occurs.
6. Opening of the large intestine.
8. First part of the small intestine, where chyme enters from the stomach.

9. Type of gland found in the stomach.

Down
1. Large, multifunction organ in the abdominal cavity, produces bile.
2. Muscular tube leading from the pharynx to the stomach.
5. Muscular sac that stores bile.
7. Lymphatic vessel inside a villus of the small intestine.

CHAPTER TEST

OBJECTIVE TEST

Do not refer to the text when taking this test.
In questions 1–6, match the statements to these digestive system structures:

    a. pharynx b. soft palate c. esophagus
    d. stomach e. small intestine f. large intestine

_____ 1. Blocks food from entering nasal passages

_____ 2. Conducts food from throat to stomach

_____ 3. Common passageway for food and air

_____ 4. Longest portion of the digestive tract

_____ 5. Contains cecum, colon, rectum, and anus

_____ 6. Has a very acidic pH

_____ 7. The layer of the digestive tract that keeps it from sticking to surrounding internal organs is the
    a. serosa.
    b. submucosa.
    c. mucosa.
    d. muscularis.

In questions 8–12, match these organs to the appropriate function.

    a. large intestine b. stomach c. liver
    d. pancreas e. gallbladder

_____ 8. Secretes digestive enzymes and sodium bicarbonate

_____ 9. Releases bile into the duodenum

____10. Absorbs water and alcohol; initiates protein breakdown
____11. Detoxifies substances removed from blood
____12. Absorbs water and houses bacteria
____13. Cholecystokinin (CCK) is a(n) _____ secreted in response to _____ in food.
   a. enzyme; protein
   b. hormone; fats
   c. lipid; carbohydrates
   d. hormone; carbohydrates
____14. Which enzyme digests starch?
   a. lipase
   b. carboxypeptidase
   c. trypsin
   d. amylase
____15. HCl
   a. is an enzyme.
   b. creates the acid environment necessary for pepsin to work.
   c. is found in the intestines.
   d. All of these are correct.
____16. Two enzymes involved in the digestion of protein are
   a. salivary amylase and lipase.
   b. trypsin and hydrochloric acid.
   c. pancreatic amylase and trypsin.
   d. pepsin and trypsin.
____17. What is absorbed into the lacteals?
   a. proteins
   b. fats
   c. carbohydrates
   d. water and amino acids
____18. Villi
   a. are found in the small intestine.
   b. increase the absorptive surface area.
   c. contain capillaries.
   d. All of these are correct.

____19. The enzymes for digestion are referred to as hydrolytic because they require
   a. hydrogen.
   b. HCl.
   c. energy.
   d. water.
____20. What type of food has the highest fat content?
   a. meat
   b. bread
   c. potatoes
   d. fruits
____21. What is the best way to ensure that you are obtaining plenty of vitamins and minerals in your diet?
   a. Take mega-multiple vitamin supplements.
   b. Be sure you eat five fruit and vegetable servings daily.
   c. Eat lots of red meat.
   d. Eat white bread.
____22. Which food group supplies quick energy?
   a. milk and cheese
   b. lipids
   c. whole-grain foods
   d. proteins
____23. An eating disorder characterized by binging on food, then inducing vomiting or other purging is
   a. bulimia nervosa.
   b. anorexia nervosa.
   c. obesity.
   d. cirrhosis.
____24. Which type of fat is best to consume in quantity?
   a. saturated fats
   b. monounsaturated fats
   c. polyunsaturated fats
   d. None of these should be consumed in quantity.
____25. Which of the following are considered antioxidants?
   a. calcium and sodium
   b. B vitamins and selenium
   c. vitamin E and iron
   d. vitamins C, E, and A

THOUGHT QUESTIONS

Answer in complete sentences.
26. Why is digestion a necessary process for humans?

27. With regard to LDL and HDL, which is "bad," and which is "good," and why?

28. How do the digestive system and reproduction interact?

Test Results: _____ number correct ÷ 28 = _____ × 100 = _____%

ANSWER KEY

STUDY QUESTIONS

1. a. mouth **b.** teeth **c.** Salivary **d.** saliva **e.** taste buds **f.** tongue **2. a.** pharynx **b.** epiglottis **3. a.** stomach **b.** peristalsis **c.** sphincter (constrictor) **4. a.** submucosa **b.** muscularis **c.** serosa **5. a.** goblet cells **b.** mucus **c.** hydrochloric (HCl) **d.** gastric **e.** pepsin **6. a.** duodenum **b.** pancreas **c.** gallbladder **d.** chyme **7. a.** villi **b.** microvilli **c.** Glucose **d.** amino acids **e.** lacteals **8.** The purpose of microvilli is to increase the surface area available for absorption in the small intestine. **9. a.** Gastrin **b.** gastric inhibitory peptide **c.** Secretin **10. a.** feces **b.** polyps **c.** fiber **11. a.** bicarbonate **b.** enzymes **12. a.** bile **b.** gallbladder **c.** gatekeeper **d.** small intestine **e.** storing iron and fat-soluble vitamins **f.** storing glycogen **g.** making blood plasma proteins **13. a.** Hepatitis **b.** cirrhosis **14. a.** Salivary amylase **b.** Mouth **c.** Basic **d.** Pancreas **e.** Maltase **f.** Small intestine **g.** Acidic **h.** Gastric glands **i.** Trypsin **j.** Small intestine **k.** Basic **l.** Small intestine **m.** Nuclease **n.** Small intestine **o.** Basic **p.** Small intestine **q.** Lipase **r.** Small intestine **15. a.** pepsin/trypsin **b.** maltase **c.** nuclease **d.** salivary amylase/pancreatic amylase **e.** lipase **f.** nucleosidases **g.** peptidase **16.** see Fig. 7.13, p. 128, in text. **17.** c **18.** d **19. a.** T **b.** T **c.** F **d.** T **20. a.** coenzymes **b.** A **c.** D **d.** C, E, and A **21. a.** sodium **b.** calcium **22. a.** B **b.** A **c.** O **d.** O **e.** A

DEFINITIONS CROSSWORD

Across:
1. lipase **3.** fiber **4.** pharynx **6.** anus **8.** duodenum **9.** gastric
Down:
1. liver **2.** esophagus **5.** gallbladder **7.** lacteal

CHAPTER TEST

1. b **2.** c **3.** a **4.** e **5.** f **6.** d **7.** a **8.** d **9.** e **10.** b **11.** c **12.** a **13.** b **14.** d **15.** b **16.** d **17.** b **18.** d **19.** d **20.** a **21.** b **22.** c **23.** a **24.** d **25.** d **26.** Like all other organisms, humans function at the level of the cell. Digestion occurs so that nutrients in foods can be processed to the point that they are available to individual cells. **27.** *LDL* refers to low-density lipoproteins. It is considered "bad" because it carries cholesterol away from the liver to the cells. *HDL* refers to high-density lipoproteins. HDL is considered "good" because it deposits cholesterol in the liver where it converts to bile salts and is eventually eliminated from the body. **28.** The digestive system provides nutrients needed for reproduction and the growth of an unborn fetus while it develops in the uterus. Later, the newborn can be supplied with nutrients from its mother during nursing, all supplied by the digestive system.

8
RESPIRATORY SYSTEM

Breathing is the first step of getting oxygen into the body so that it can eventually be used by your cells in the manufacture of ATP. In this chapter, you will learn how respiration is controlled by your brain (p. 149) and about the organs involved in this life-giving function (pp. 142–145).

Remember that the term *respiration* in biology means two different things. Cellular respiration is the metabolic pathway undertaken by your cells to break down glucose and transfer energy to ATP molecules. In this chapter, you will see that *respiration* also refers to the mechanisms of breathing.

When you study the material in this chapter, be sure to differentiate between external and internal respiration (p. 150). External respiration is the exchange of gases between the air inside the lungs and the blood, and internal respiration is exchange occurring between the blood and tissue cells. Make a detailed, step-by-step summary of the ways in which hemoglobin participates in these two processes (p. 150).

How environmental pollutants, including cigarette smoking (pp. 153–154), and diseases (pp. 152–154) damage your lungs is also discussed. Once you realize how a thin alveolar membrane facilitates rapid gas exchange, it will become clear why alveoli are so easily damaged.

The respiratory system contributes to homeostasis by ridding blood of CO_2 and making it O_2-rich.

TIP: Be sure to visit the Online Learning Center that accompanies *Human Biology* 9/e. It has practice quizzes, interactive activities, labeling exercises, art quizzes, animation, flash cards, and much more. http://www.mhhe.com/maderhuman9

Study the text section by section. Answer the study questions so that you can fulfill the learning objectives for each section.

8.1 THE RESPIRATORY SYSTEM (PAGES 142–145)

After you have answered the questions for this section, you should be able to
- Describe the pathway air takes in and out of the lungs and the structures involved that are designed to filter, warm, and moisten air.

1. Complete this table. Refer to Table 8.1 in the textbook as needed.

Structure	Function
Nares	a. _____
b. _____	Filter, warm, and moisten air
c. _____	Connection to surrounding regions
Glottis	d. _____
e. _____	Sound production
Trachea	f. _____
g. _____	Passage of air to lungs
Bronchioles	h. _____
i. _____	Contains alveoli and blood vessels

2. Label the following diagram using the alphabetized list of terms.
   epiglottis
   glottis
   hard palate
   larynx
   nasal cavity
   soft palate
   trachea

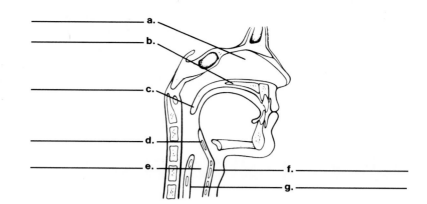

a. _____
b. _____
c. _____
d. _____
e. _____
f. _____
g. _____

3. The nasal cavities contain ᵃ._____ for smelling, and each, as well as the trachea, is lined with ᵇ._____ to screen the incoming air. During swallowing, the ᶜ._____ folds down over the glottis to keep food from entering the trachea. The lungs of premature infants often lack a film, called ᵈ._____, which keeps their lung tissues from sticking together.

8.2 MECHANISM OF BREATHING (PAGES 146–149)

After you have answered the questions for this section, you should be able to
- Relate the respiratory volumes to a diagram showing the amount of air that is moved in and out of the lungs when breathing occurs.
- Describe the mechanism by which breathing occurs, including inspiration and expiration.

4. Label the following diagram using the alphabetized list of terms.
   expiratory reserve volume
   inspiratory reserve volume
   residual volume (used twice)
   tidal volume

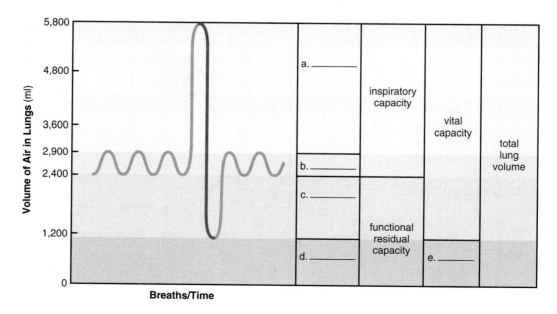

5. In the preceding diagram, the sum of the volumes labeled *a, b,* and *c* is termed the _____.

6. Place the appropriate letter next to each phrase.

   I—inspiration   E—expiration

   a. _____ lungs expanded
   b. _____ muscles (diaphragm and ribs) relaxed
   c. _____ diaphragm dome-shaped
   d. _____ chest enlarged
   e. _____ less air pressure in lungs than in the environment

7. What is the proper sequence for these statements? (Indicate by letters.)
   a. Respiratory center stops sending nerve impulse to diaphragm and rib cage.
   b. Respiratory center sends nerve impulse to diaphragm and rib cage.
   c. Diaphragm relaxes and becomes dome-shaped, and rib cage moves down and inward.
   d. Lungs expand as diaphragm lowers, and rib cage moves upward and outward.
   e. Air goes rushing out as lungs recoil.
   f. Air comes rushing in as lungs expand.

8.3 GAS EXCHANGES IN THE BODY (PAGES 150–151)

After you have answered the questions for this section, you should be able to
• Describe the events that occur during external and internal respiration.

8. Match the statements to these terms:

   internal respiration   cellular respiration   inspiration and expiration   external respiration

   a. _____ entrance and exit of air into and out of lungs
   b. _____ exchange of gases between blood and tissue fluid
   c. _____ production of ATP in cells
   d. _____ exchange of gases between lungs and blood
   e. Next, place the terms in the proper sequence.

   First _____

   Second _____

   Third _____

   Last _____

9. Give the equation that describes how oxygen is transported in the blood. Label one arrow *lungs* and the reverse arrow *tissues*.

10. a. Give the equation that describes how most of the carbon dioxide is transported in the blood. Label one arrow *lungs* and the reverse arrow *tissues*.

    b. What is the name of the enzyme that speeds this reaction? _____

    c. Carbon dioxide transport produces hydrogen ions. Why does the blood not become acidic? _____

    _____

    d. By what process does carbon dioxide move from the blood to the alveoli? _____

11. After studying Figure 8.9 in the text, fill in the blanks.
    a. Where does oxygen enter the blood? _____
    b. Where does oxygen exit the blood? _____
    c. Where does carbon dioxide enter the blood? _____
    d. Where does carbon dioxide exit the blood? _____

e. What two types of vessels are high in oxygen?

_____

f. What two types of vessels are high in carbon dioxide? _____

_____

8.4 RESPIRATION AND HEALTH (PAGES 152–154)

After you have answered the questions for this section, you should be able to
- List the names, symptoms, and causes of various diseases of the respiratory tract.
- Explain why there is an increase of lung cancer in women.

12. Match the descriptions to these terms:

    tonsillitis   pneumonia   tuberculosis   emphysema   pulmonary fibrosis   lung cancer

a. _____ Cells build a protective capsule around the bacteria. X rays can detect the presence of these capsules.

b. _____ Fibrous connective tissue builds up in the lungs of a person who has inhaled particles.

c. _____ A first line of defense against an invasion of the body by pathogens.

d. _____ This most often begins in a bronchus and is caused by smoking cigarettes.

e. _____ Lungs balloon because air is trapped in the alveoli.

f. _____ Lobules of the lungs fill with fluid; infection is caused by a microbe.

13. Why do women now suffer from increased lung cancer rates?

_____

DEFINITIONS WORDSEARCH

Review key terms by completing this wordsearch using the following alphabetized list of terms:

```
A G L O T T I S A G A I
E X I A L V E O L U S N
P P T O R S T O P T O S
Y G I T Q Y B E G U N U
T L D G R E N L I V I R
I D A S L C I X I S N F
C A L V W O J H R O O A
A D V I I L T C I L I C
P G O D L L A T O X T T
A A L A V A A K I E A A
C V V G R L E I V S R N
L D M O I S V E M Y I T
A G E T A Y E R L B P V
T B N J J N C M P O X M
I E I M E N H A I R E I
V O C A L C O R D O O N
```

alveolus
epiglottis
expiration
glottis
larynx
tidal volume
ventilation
vital capacity
vocal cord

a. _____ Contains the vocal cords.
b. _____ Opening for airflow into the larynx.
c. _____ Act of expelling air.
d. _____ Amount of air involved in normal inhale/exhale cycle.
e. _____ Process of breathing.
f. _____ Fold of tissue in larynx that creates sounds.
g. _____ Structure that covers glottis during swallowing.
h. _____ Air sac in the lung.
i. _____ Maximum amount of air moved in to or out of lungs during breathing.

CHAPTER TEST

OBJECTIVE TEST

Do not refer to the text when taking this test.

____ 1. Why is oxygen needed by the body?
   a. to aerate the lungs
   b. to cleanse the blood
   c. to produce ATP
   d. Both *a* and *b* are correct.

____ 2. The structure(s) that receive(s) air after the trachea is(are) the
   a. pharynx.
   b. bronchi.
   c. bronchiolus.
   d. alveoli.

____ 3. Which structure carries both air and food?
   a. larynx
   b. pharynx
   c. trachea
   d. esophagus

____ 4. Which of the following contains only parts of the upper respiratory tract?
   a. lungs, larynx, bronchi
   b. nasal cavities, alveoli, trachea
   c. lungs, alveoli, bronchi
   d. nasal cavities, pharynx, larynx

____ 5. Which of these constricts during an asthma attack?
   a. trachea
   b. bronchus
   c. bronchiole
   d. pharynx

____ 6. The alveoli
   a. are sacs in the lungs.
   b. contain capillaries.
   c. are where gas exchange occurs.
   d. All of these are correct.

____ 7. Which of these contains the vocal cords?
   a. glottis
   b. epiglottis
   c. pharynx
   d. larynx

____ 8. Before oxygen is picked up in the lungs by hemoglobin, it first diffuses through (a) alveolar cells, (b) blood plasma, (c) red blood cell plasma membranes, and (d) capillary walls, though not necessarily in this order. What is the correct order?
   a. a, b, d, c
   b. a, d, b, c
   c. d, a, c, b
   d. d, b, a, c
   e. a, b, c, d

____ 9. When the lungs recoil,
   a. inspiration occurs.
   b. external respiration occurs.
   c. internal respiration occurs.
   d. expiration occurs.
   e. All of these are correct.

____ 10. The respiratory center
   a. is stimulated by hydrogen ions.
   b. is located in the chest.
   c. sends nerve impulses to lung tissue.
   d. is stimulated by oxygen levels.

____ 11. The amount of air that enters or leaves the lungs during a normal respiratory cycle is the
   a. tidal volume.
   b. respiratory volume.
   c. residual volume.
   d. vital capacity.

____ 12. The maximum amount of air a person can exhale after taking the deepest breath possible is a measure of the
   a. residual volume.
   b. tidal volume.
   c. vital capacity.
   d. inspiratory reserve volume.

____ 13. External respiration is defined as
   a. an exchange of gases in the lungs.
   b. breathing.
   c. an exchange of gases in the tissues.
   d. cellular respiration.

_____ 14. Which gas is primarily carried by the plasma?
  a. O_2
  b. CO_2
  c. both O_2 and CO_2
  d. neither O_2 nor CO_2

_____ 15. CO_2 enters the blood as a result of
  a. active transport.
  b. diffusion.
  c. blood pressure.
  d. air pressure.

_____ 16. The enzyme carbonic anhydrase causes
  a. carbon dioxide to react with water.
  b. carbon dioxide to react with bicarbonate ions.
  c. water to react with hydrogen ions.
  d. Both _b_ and _c_ are correct.

_____ 17. Hemoglobin combines with
  a. oxygen more readily in the lungs.
  b. carbon dioxide more readily in the tissues.
  c. oxygen more readily in the tissues.
  d. carbon dioxide more readily in the lungs.
  e. Both _a_ and _b_ are correct.

_____ 18. Hemoglobin carries
  a. O_2.
  b. CO_2.
  c. hydrogen ions.
  d. All of these are correct.

_____ 19. Which lung disorder is NOT caused by a pathogen?
  a. pneumonia
  b. tuberculosis
  c. emphysema
  d. laryngitis

_____ 20. Smoking cigarettes
  a. causes tuberculosis.
  b. leads to emphysema and cancer.
  c. increases the vital capacity of the lungs.
  d. leads to good health and longer life.

In questions 21–23, match the statements to these items:
  a. pneumonia   b. lung cancer   c. infant respiratory distress

_____ 21. Nonfunctional tissues interfere with gas exchange

_____ 22. Fluid-filled aveoli

_____ 23. Alveolar collapse due to high surface tension

_____ 24. Which body system does the respiratory system aid by providing oxygen so neurons can function properly?
  a. nervous system
  b. lymphatic system
  c. cardiovascular system
  d. integumentary system

_____ 25. Which body system helps the respiratory system by protecting the lungs and providing points for breathing muscle attachment?
  a. cardiovascular system
  b. skeletal system
  c. muscular system
  d. urinary system

THOUGHT QUESTIONS

Answer in complete sentences.

26. Explain how expiration occurs once the lungs have filled with air.

27. Relate the large surface area provided by the alveoli to the process by which external respiration occurs.

Test Results: _____ number correct ÷ 27 = _____ × 100 = _____%

ANSWER KEY

STUDY QUESTIONS

1. a. Passage of air into nasal cavities **b.** Nasal cavities **c.** pharynx **d.** Passage of air into larynx **e.** Larynx **f.** Passage of air to bronchi **g.** Bronchi **h.** Passage of air to each alveolus **i.** Lungs   **2. a.** nasal cavity **b.** hard palate **c.** soft palate **d.** epiglottis **e.** glottis **f.** larynx **g.** trachea   **3. a.** odor receptors **b.** cilia **c.** epiglottis **d.** surfactant   **4. a.** inspiratory reserve volume **b.** tidal volume **c.** expiratory reserve volume **d.** residual volume **e.** residual volume   **5.** vital capacity   **6. a.** I **b.** E **c.** E **d.** I **e.** I   **7.** b, d, f, a, c, e   **8. a.** inspiration and expiration **b.** internal respiration **c.** cellular respiration **d.** external respiration **e.** inspiration and expiration, external, internal, cellular

9.
$$Hb + O_2 \underset{tissues}{\overset{lungs}{\rightleftharpoons}} HbO_2$$

10. a.
$$CO_2 + H_2O \underset{lungs}{\overset{tissues}{\rightleftharpoons}} H_2CO_3 \underset{lungs}{\overset{tissues}{\rightleftharpoons}} H^+ + HCO_3^-$$

b. carbonic anhydrase **c.** Hemoglobin combines with excess hydrogen ions. **d.** diffusion   **11. a.** lungs **b.** tissues **c.** tissues **d.** lungs **e.** pulmonary vein and systemic arteries **f.** systemic veins and pulmonary artery   **12. a.** tuberculosis **b.** pulmonary fibrosis **c.** tonsillitis **d.** lung cancer **e.** emphysema **f.** pneumonia   **13.** Increased numbers of women who smoke.

DEFINITIONS WORDSEARCH

```
    G L O T T I S
  E     A L V E O L U S
    P T   R
  Y   I       Y
  T   D G       N
  I   A   L     X     N
  C   L       O     O O
  A   V       T   I   I
  P   O         T   T
  A   L       A   I   A
  C   V       L     S R
  L   M   I           I
  A   E             P
  T   N             X
  I E               E
  V O C A L C O R D
```

a. larynx **b.** glottis **c.** expiration **d.** tidal volume **e.** ventilation **f.** vocal cord **g.** epiglottis **h.** alveolus **i.** vital capacity

CHAPTER TEST

1. c **2.** b **3.** b **4.** d **5.** c **6.** d **7.** d **8.** b **9.** d **10.** a **11.** a **12.** c **13.** a **14.** b **15.** b **16.** b **17.** e **18.** d **19.** c **20.** b **21.** b **22.** a **23.** c **24.** a **25.** b   **26.** The diaphragm relaxes when the respiratory center stops sending messages to contract. Once relaxation occurs, expiration is passive. Air leaves the lungs with the elastic recoil of the lungs. Muscle contraction can force additional air from the lungs.   **27.** Since oxygen enters the capillaries of the alveoli by the process of diffusion, a passive process, a large surface area is required.

9

URINARY SYSTEM AND EXCRETION

The human kidney is an intriguing structure designed to conserve water and to excrete a concentrated urine. Unless you take the time to study the topics of this chapter carefully, it is easy to become overwhelmed by the variety of events occurring in the nephron, the functional unit of the kidney.

Begin your study of this chapter by first reviewing the structures involved in the urinary system (p. 160). Learn the macroscopic structure of the kidney (p. 163); then go inside it and learn about the nephron. Draw and label a diagram of a nephron and add the path of blood about the nephron (pp. 164–165). Next, associate the three steps in urine formation to parts of the

nephron and tell how each part is suited to its function (pp. 166–167). Explain the medical problems that arise when kidneys do not function properly. Pay particular attention to the process of hemodialysis.

Now you are ready to study how the kidneys contribute to homeostasis by maintaining water-salt balance (pp. 168–169) and maintaining acid-base balance (p. 170). Again, associate these functions with the parts of the nephron. Also identify the role of particular hormones. Study the illustration on page 173 that shows how the other systems of the body work with the urinary system to maintain homeostasis.

TIP: Be sure to visit the Online Learning Center that accompanies *Human Biology* 9/e. It has practice quizzes, interactive activities, labeling exercises, art quizzes, animations, flash cards, and much more. http://www.mhhe.com/maderhuman9

STUDY QUESTIONS

Study the text section by section. Answer the study questions so that you can fulfill the learning objectives for each section.

9.1 URINARY SYSTEM (PAGES 160–161)

After you have answered the questions for this section, you should be able to
- List the functions of the urinary system that maintain homeostasis.
- Trace the path of urine, and describe the general structure and function of each organ mentioned.
- Describe how urination is controlled by the nervous system.

1. As the kidneys produce urine, what four functions help maintain homeostasis?

   a. _____

   b. _____

   c. _____

   d. _____

2. Match the functions to these urinary organs:

   kidney   ureter   urinary bladder   urethra

   a. _____ muscular tube leading from kidneys to urinary bladder

   b. _____ tube leading from bladder to the outside

   c. _____ hollow, muscular organ that stores urine

   d. _____ bean-shaped organ that filters blood

3. What triggers urination? _____
_____
_____

9.2 KIDNEY STRUCTURE (PAGES 163–165)

After you have answered the questions for this section, you should be able to
- Describe the macroscopic structure of the kidney.
- State the parts of a nephron and relate these to the macroscopic anatomy.

4. Macroscopically, the kidney is composed of these three parts:

a. _____

b. _____

c. _____

5. Label the following diagram of the parts of the nephron, using the following alphabetized list of terms.

afferent arteriole	efferent arteriole	loop of the nephron
collecting duct	glomerular capsule	peritubular capillary network
distal convoluted tubule	glomerulus	proximal convoluted tubule

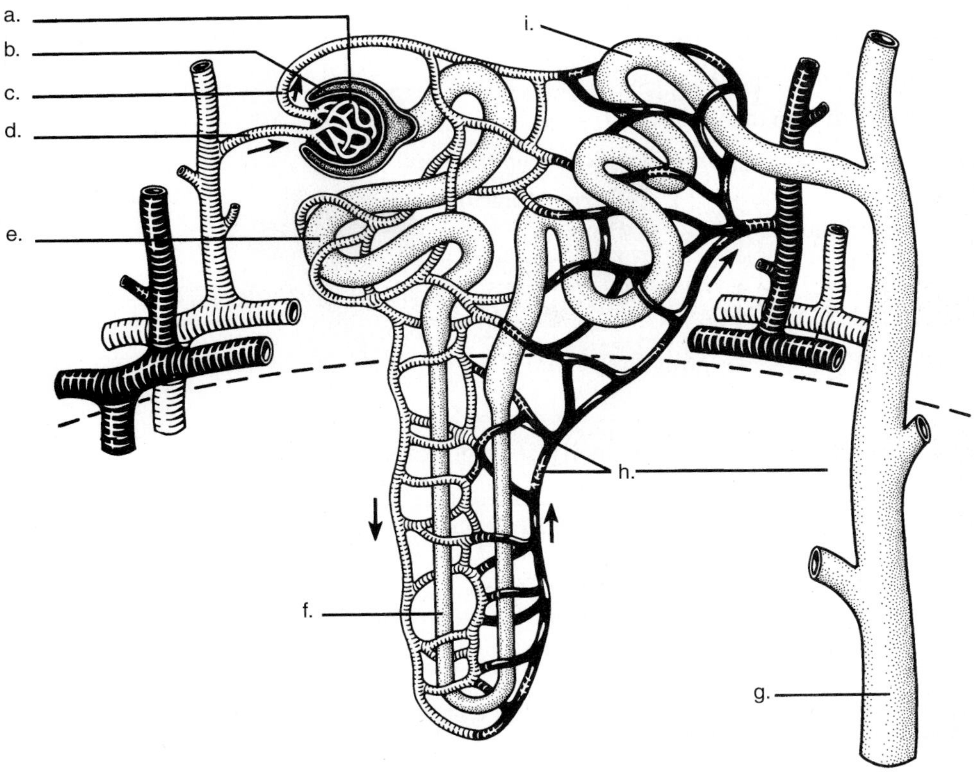

a. _____
b. _____
c. _____
d. _____
e. _____
f. _____
g. _____
h. _____
i. _____

6. Trace the path of filtrate from the glomerular capsule to the collecting duct.
glomerular capsule

a. _____

b. _____

c. _____
collecting duct

After you have answered the questions for this section, you should be able to
• Describe the three steps in urine formation, and relate these to parts of a nephron.

7. In the following diagram, add the three steps in urine formation.

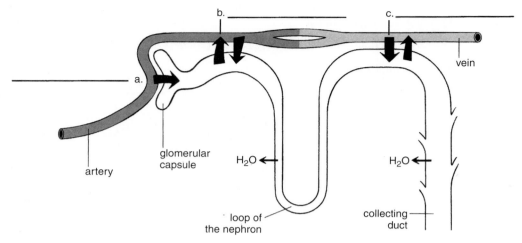

d. Where does the first step take place? _____

e. Where does the second step take place? _____

f. Where does the third step take place? _____

8. Blood in the glomerulus is composed of these two portions:

Small Molecules	Large Molecules, etc.
Nutrients (glucose, amino acids)	Proteins
Wastes (urea, uric acid)	Formed elements
Salts	
Water	

a. Which portion will undergo filtration?

_____

b. Which of the small molecules will maximally undergo reabsorption?

_____

c. Which of the small molecules will minimally undergo reabsorption?

_____

9. Give an example of a molecule that undergoes tubular secretion. _____

_____

9.4 REGULATORY FUNCTIONS OF THE KIDNEYS (PAGES 168–170)

After you have answered the questions for this section, you should be able to
• Describe how the loop of the nephron contributes to water reabsorption.
• Name three hormones involved in maintaining blood volume, and explain how they function.

10. The presence of which parts of the nephron accounts for maximal reabsorption of water in humans?

a. _____ What causes water to leave these parts of the nephron?

b. _____

Because water is maximally reabsorbed, humans excrete a c. _____ urine.

11. Complete this table.

ADH	Urine Quantity
Increased amount	a.
Reduced amount	b.

12. Fill in the blanks with these terms:

blood pressure rises   adrenal cortex   converting enzyme   renin   aldosterone   atrial natriuretic hormone
angiotensin II

a. _____ changes angiotensinogen to angiotensin I. Converting enzyme changes

angiotensin I to b. _____ Angiotensin II acts on the c. _____

to secrete d. _____. Kidneys absorb Na$^+$ and blood pressure

e. _____. When blood pressure rises, the heart secretes f. _____, which

causes the kidneys to excrete Na$^+$.

13. If the blood is acidic, a. _____ ions are excreted in combination with

b. _____, while c. _____ are reabsorbed. If the blood is basic,

fewer d. _____ ions are excreted, and fewer e. _____ are reabsorbed.

9.5 PROBLEMS WITH KIDNEY FUNCTION (PAGE 171)

After you have answered the questions for this section, you should be able to
* List the various illnesses of the urinary tract, and discuss the available treatments.

14. Both of the following indicate a problem with kidney function. Why?

a. albumin in the urine _____

b. high blood urea _____

15. The simplified diagram below explains how the artificial kidney works.

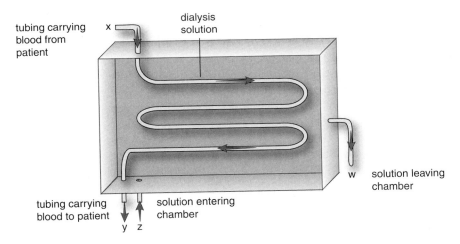

At x, the patient's blood carries impurities. As it passes through the artificial kidney (box), the impurities pass out
of the blood. At y, the patient's blood is rid of nitrogenous wastes. What should be the makeup of the solution
that enters the chamber at z? _____

_____

9.6 HOMEOSTASIS (PAGES 172–173)

After you have answered the questions for this section, you should be able to
- Explain how the activities of the urinary system and other body systems work together to maintain homeostasis.
- Describe how the kidneys have ultimate control over the pH of the blood.

16. The *Human System Work Together* diagram in your textbook indicates ways in which the urinary system benefits other organ systems of the body. Match the organ systems with the correct descriptions.
    (1) Kidneys control volume of body fluids, including lymph.
    (2) Kidneys maintain blood levels of Na^+, K^+, and Ca^{2+}, needed for muscle innervation, and eliminate creatinine, a muscle waste.
    (3) Kidneys work with lungs to maintain blood pH.
    (4) Kidneys convert vitamin D to active form needed for Ca^{2+} absorption and compensate for any water loss by digestive tract.
    (5) Kidneys maintain blood levels of Na^+, K^+, and Ca^{2+}, needed for nerve conduction.
    (6) Kidneys keep blood values within normal limits so that transport of hormones continues.
    (7) Kidneys filter blood and excrete wastes; maintain blood volume, pressure, and pH; and produce renin and erythropoietin.
    (8) Semen is discharged through the urethra in males; kidneys excrete wastes and maintain electrolyte levels for mother and child.
    (9) Kidneys compensate for water loss due to sweating and activate vitamin D precursor made by skin.

    _____ a. integumentary system
    _____ b. lymphatic system/immunity
    _____ c. cardiovascular system
    _____ d. muscular system
    _____ e. nervous system
    _____ f. respiratory system
    _____ g. reproductive system
    _____ h. endocrine system
    _____ i. digestive system

In the table, place an X beside the component of blood if the following descriptions pertain to it:

a. in the afferent arteriole
b. in the filtrate
c. in the efferent arteriole
d. reabsorbed into the peritubular capillary
e. secreted from the peritubular capillary
f. present in urine
g. absent from urine
h. in venous blood

	a	b	c	d	e	f	g	h
1. plasma proteins	X		X				X	X
2. red blood cells	X		X				X	X
3. white blood cells	X		X				X	X
4. glucose	X	X		X			X	
5. amino acids	X	X		X		X		
6. sodium chloride	X	X		X		X		X
7. water	X	X		X		X		X
8. urea	X	X				X		X
9. uric acid	X	X				X		
10. penicillin	X	X	X		X	X		

There are 45 correct answers. There are 80 possible errors of omission or commission. Any 10 errors and you're ELIMINATED!

DEFINITIONS CROSSWORD

Review key terms by completing this crossword puzzle using the following alphabetized list of terms:

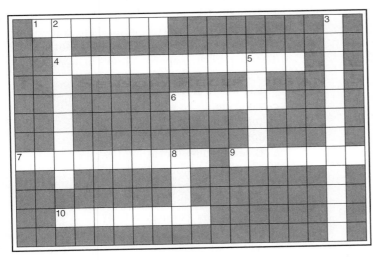

antidiuretic
collecting duct
excretion
glomerulus
kidney
nephron
proximal
urea
ureter
urethra

Across
1 Anatomical and functional unit of the kidney.
4 Tube that receives urine from the distal convoluted tubules of several nephrons.
6 Organ of the urinary system that produces and excretes urine.
7 Ball of capillaries, surrounded by a glomerular capsule of a nephron.
9 Tube conveying urine from the bladder to outside the body.
10 Convoluted tubule nearest the glomerular capsule.

Down
2 Process of removing metabolic wastes from the body.
3 Type of hormone from the posterior pituitary that promotes the reabsorption of water from the collecting duct.
5 One of two tubes leading from the kidneys to the urinary bladder.
8 Main nitrogenous waste derived from the breakdown of amino acids.

CHAPTER TEST

OBJECTIVE TEST

Do not refer to the text when taking this test.

____ 1. Kidneys are organs of homeostasis because they
   a. regulate the blood volume.
   b. regulate the pH of the blood.
   c. help maintain the correct concentration of salts in the blood.
   d. excrete nitrogenous wastes.
   e. All of these are correct.

____ 2. Which of these contains urine?
   a. urethra
   b. uterus
   c. intestine
   d. gallbladder
   e. All of these are correct.

____ 3. Which portion of the urinary tract varies significantly in length or size between males and females?
   a. kidneys
   b. ureters
   c. urinary bladder
   d. urethra

____ 4. Urination is triggered by
   a. contraction of the bladder and relaxation of sphincter muscles.
   b. relaxation of the bladder and contraction of sphincter muscles.
   c. contraction of kidney and ureter muscles.
   d. contraction of kidneys and relaxation of the urinary bladder.

____ 5. Which portion is NOT part of the kidney?
   a. renal cortex
   b. renal urethra
   c. renal medulla
   d. renal pelvis

____ 6. The collecting ducts are primarily in the
   a. renal cortex.
   b. renal medulla.
   c. renal pelvis.
   d. afferent arteriole.

In questions 7–11, match the description with these structures:
   a. glomerulus   b. glomerular capsule
   c. renal cortex   d. loop of the nephron
   e. collecting duct

____ 7. Often extends into the medulla
____ 8. A knot of capillaries
____ 9. Variably permeable to water
____ 10. Site of afferent/efferent arterioles
____ 11. Blind end of the proximal convoluted tubule

____ 12. Urine collects in the _____ before entering the ureter.
   a. renal medulla
   b. renal cortex
   c. renal pelvis
   d. capsule

____ 13. Glomerular filtration should be associated with
   a. the glomerular capsule.
   b. the distal convoluted tubule.
   c. the collecting duct.
   d. All of these are correct.

____ 14. Sodium is removed from the nephron by
   a. passive and active reabsorption.
   b. tubular secretion.
   c. an attraction to Cl^-.
   d. secretion.

____ 15. Tubular reabsorption occurs primarily at
   a. the glomerular capsule.
   b. the proximal convoluted tubule.
   c. the loop of the nephron.
   d. the distal convoluted tubule.

____ 16. In humans, water is
   a. found in the glomerular filtrate.
   b. reabsorbed from the nephron.
   c. in the urine.
   d. All of these are correct.

____ 17. Glucose
   a. is in the filtrate and urine.
   b. is in the filtrate but not in urine.
   c. undergoes tubular secretion and is in urine.
   d. undergoes tubular secretion but is not in urine.

____ 18. The loop of the nephron allows us to excrete
   a. a diluted urine.
   b. a concentrated urine.
   c. no urine.
   d. too much urine.

____ 19. Aldosterone
   a. is secreted by the adrenal cortex.
   b. causes the blood volume to lower.
   c. is the same as renin.
   d. causes the kidneys to excrete sodium.

____ 20. Which of these is mismatched?
   a. aldosterone—adrenal cortex
   b. renin—kidneys
   c. ADH—anterior pituitary
   d. ANH—heart

____ 21. ADH is necessary for
   a. water reabsorption.
   b. glucose reabsorption.
   c. protein reabsorption.
   d. All of these are correct.

_____22. If the nephrons do not function,
  a. urea accumulates in the blood.
  b. edema occurs.
  c. water and salt balances are disturbed.
  d. All of these are correct.
_____23. The juxtaglomerular apparatus
  a. secretes renin.
  b. is sensitive to blood pressure.
  c. occurs where the afferent arteriole and distal convoluted tubule touch.
  d. has a regulatory function.
  e. All of these are correct.
_____24. Antidiuretic hormone acts directly on the collecting duct to
  a. reabsorb water and make the urine less concentrated.
  b. excrete water and make the urine more concentrated.
  c. reabsorb water and make the urine more concentrated.
  d. excrete water and make the urine less concentrated.
  e. None of these is correct.
_____25. With regard to the artificial kidney, which of these is not a proper match? Dialysis fluid
  a. contains more salt than blood—less water is withdrawn from blood.
  b. does not contain urea—more urea is withdrawn from blood.
  c. has a pH counter to blood—blood will adjust its pH.
  d. is pure—blood will give up toxic substances.

THOUGHT QUESTIONS

Answer in complete sentences.
26. Explain the difference between defecation and excretion.

27. What role is played by the high concentration of salt and urea in the renal medulla?

Test Results: _____ number correct ÷ 27 = _____ × 100 = _____%

ANSWER KEY

STUDY QUESTIONS

1. a. maintenance of salt-water balance **b.** secretion of hormones **c.** maintenance of acid-base balance (blood pH) **d.** excretion of metabolic wastes, including nitrogenous wastes **2. a.** ureter **b.** urethra **c.** urinary bladder **d.** kidney **3.** Stretch receptors send impulses to the spinal cord, which sends nerve impulses back to muscles controlling the urinary bladder. Contraction of the bladder and relaxation of sphincter muscles occur, expelling urine to the outside. **4. a.** renal cortex **b.** renal medulla **c.** renal pelvis **5. a.** glomerulus **b.** glomerular capsule **c.** efferent arteriole **d.** afferent arteriole **e.** proximal convoluted tubule **f.** loop of the nephron **g.** collecting duct **h.** peritubular capillary network **i.** distal convoluted tubule **6. a.** proximal convoluted tubule **b.** loop of the nephron **c.** distal convoluted tubule **7. a.** glomerular filtration **b.** tubular reabsorption **c.** tubular secretion **d.** between the glomerulus and the glomerular capsule **e.** primarily at the proximal convoluted tubule **f.** at the proximal convoluted tubule and the distal convoluted tubule **8. a.** small molecules **b.** nutrients, salts, water **c.** wastes **9.** hydrogen ions, penicillin, creatinine **10. a.** loop of the nephron and the collecting duct **b.** hypertonic medulla due to presence of salt (from ascending limb of loop of the nephron) and urea (from collecting duct) **c.** hypertonic **11. a.** little urine **b.** much urine **12. a.** renin **b.** angiotensin II **c.** adrenal cortex **d.** aldosterone **e.** rises **f.** atrial natriuretic hormone **13. a.** hydrogen **b.** ammonia **c.** bicarbonate ions **d.** hydrogen **e.** bicarbonate ions **14. a.** Normally, albumin does not filter out of glomerulus. Increased permeability indicates kidney failure. **b.** Normally, urea filters out of glomerulus. **15.** The solution should contain those substances (in proper concentration) that you do not wish to leave in the blood; it should lack those substances you do wish to leave in the blood. **16. a.** (9) **b.** (1) **c.** (7) **d.** (2) **e.** (5) **f.** (3) **g.** (8) **h.** (6) **i.** (4)

	a	b	c	d	e	f	g	h
plasma proteins	X		X				X	X
red blood cells	X		X				X	X
white blood cells	X		X				X	X
glucose	X	X		X			X	X
amino acids	X	X		X			X	X
sodium chloride	X	X		X		X		X
water	X	X	X	X		X		X
urea	X	X				X		X
uric acid	X	X			X	X		
penicillin	X		X		X	X		

DEFINITIONS CROSSWORD

Across:
1. nephron   4. collecting duct   6. kidney   7. glomerulus   9. urethra   10. proximal

Down:
2. excretion   3. antidiuretic   5. ureter   8. urea

CHAPTER TEST

1. e   2. a   3. d   4. a   5. b   6. b   7. d   8. a   9. e
10. c   11. b   12. c   13. a   14. a   15. b   16. d
17. b   18. b   19. a   20. c   21. a   22. d   23. e
24. c   25. a   26. Defecation is the elimination of nondigested material from the gut, and excretion is the elimination of end products of metabolism by the kidneys.
27. The high concentration of salt and urea in the renal medulla draws water out of the loop of the nephron and the collecting duct.

PART 3 MOVEMENT AND SUPPORT IN HUMANS
10
SKELETAL SYSTEM

The skeletal system protects and supports the organs of the body. It also provides points of attachment for muscles and contributes to homeostasis by storing minerals and producing blood cells for the body.

Begin your study of the skeletal system by learning about the structure of bone (pp. 178–179) and the types of tissues involved (pp. 178–179). Sketch your own diagram of a long bone (p. 179), and label the different portions.

Bone formation and remodeling are continuous throughout a lifetime. Make a chart with two columns in which you can compare the two types of bone formation,

intramembranous ossification (p. 180) and endochondral ossification (p. 181). Also make lists of the steps involved in bone remodeling (pp. 182–183) and in bone repair (p. 183).

The best way to learn the names of the skeletal bones (pp. 185–191) is to study them repeatedly and often. Refer to a clear diagram that labels bones, and quiz yourself as you learn them. Pronounce their names out loud. Saying and hearing a name while reading it will reinforce it in your memory. Review the types of joints (p. 192) and joint movements (pp. 192–193).

TIP: Be sure to visit the Online Learning Center that accompanies *Human Biology* 9/e. It has practice quizzes, interactive activities, labeling exercises, art quizzes, animations, flash cards, and much more. http://www.mhhe.com/maderhuman9

STUDY QUESTIONS

Study the text section by section. Answer the study questions so that you can fulfill the learning objectives for each section.

10.1 TISSUES OF THE SKELETAL SYSTEM (PAGES 178–179)

After you have answered the questions for this section, you should be able to
- Describe the bone, cartilage, and fibrous connective tissues that constitute the skeleton.
- Sketch and label the structure of a long bone.

1. Compact bone is composed of bone cells arranged in concentric circles called a._____. Inside compact bone, a lighter type of bone, called b._____, has spaces often filled with c._____, the blood-forming tissue. Ligaments, which attach bone to bone, are made up of d._____ tissue.

2. Name the three types of cartilage and where they can be found.

   a. _____

   b. _____

   c. _____

3. Label the following diagram of a long bone using the alphabetized list of terms.
   compact bone
   hyaline cartilage
   medullary cavity
   spongy bone

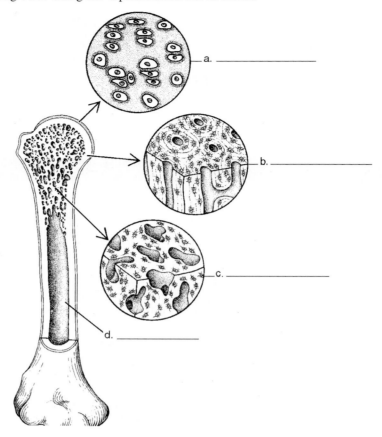

a. _____

b. _____

c. _____

d. _____

4. Which part of the long bone shown in the previous illustration is associated with red bone marrow?
   a._____ Which part of bone is the hardest? b._____
   Which part of bone is the most flexible? c._____ Which part of a long bone is associated with
   yellow bone marrow? d._____

10.2 BONE GROWTH, REMODELING, AND REPAIR (PAGES 180—183)

After you have answered the questions for this section, you should be able to
- List the different cell types of bone, and give their functions.
- Differentiate between intramembranous and endochondral ossification.
- Understand the function of the epiphyseal plates of developing long bones.
- Explain the process of remodeling of bones.
- List the sequence of events that occur as a bone repairs a fracture.

5. a. What is the relationship between osteoblasts and osteocytes? _____
   b. What is the function of an osteoclast? _____
   c. What is the process of bone formation called? _____
   d. What role is played by the growth plates of long bones? _____

6. Place the appropriate letter next to each statement. I—intramembranous ossification E—endochondral ossification
   a._____ skull bones
   b._____ cartilaginous models of bones
   c._____ epiphyseal plates remain as the bones grow in length
   d._____ spongy bone laid down first, compact bone forms over top
   e._____ bones develop between sheets of fibrous connective tissue

7. During the continual process of bone remodeling, osteoclasts remove worn-out bone cells. At the same time,

   a._____ is released into the bloodstream. The two factors that affect bone thickness are

   b._____ and c._____.

8. Repair of fractures. Place the following steps in the proper sequence by rearranging the letters.

   a._____ fibrocartilage callus

   b._____ bony callus

   c._____ remodeling

   d._____ hematoma forms

10.3 BONES OF THE SKELETON (PAGES 185–191)

After you have answered the questions for this section, you should be able to
- List the functions of the skeleton.
- Identify and state a function for the bones of the axial skeleton, including the cranium and face.
- Identify and state a function for the bones of the appendicular skeleton.

9. Name five functions of the skeleton.

   a._____

   b._____

   c._____

   d._____

   e._____

10. Axial versus appendicular skeleton. Write *ax* in front of all bones belonging to the axial skeleton; write *ap* in front of all bones belonging to the appendicular skeleton. Write *pec* in front of all bones belonging to the pectoral girdle; write *pel* in front of all bones belonging to the pelvic girdle. Some items have more than one answer.

   a._____ coxal bone   g._____ ribs

   b._____ sternum   h._____ radius

   c._____ humerus   i._____ clavicle

   d._____ scapula   j._____ tibia

   e._____ skull   k._____ fibula

   f._____ femur   l._____ ulna

11. The a._____ bone forms the forehead, and the b._____ bone has the foramen magnum, through which the spinal cord passes. The c._____ bones have an opening for the ears. The d._____ is the only movable portion of the skull and permits us to chew our food. The e._____ form the upper jaw and anterior hard palate. The f._____ bones form the cheekbone, and the nasal bones form the bridge of the nose.

12. a. Give a function of vertebrae. _____

    b. Give the name of the vertebrae that have ribs. _____

    c. Give the name of the vertebrae in the lower back. _____

    d. Give the name of the vertebrae in the neck. _____

13. Comparisons. The radius and ulna are to the forearm as the a._____ and b._____ are to the leg. The femur is to the thigh as the c._____ is to the upper arm. The metacarpals are to the palm as the d._____ are to the foot.

14. Give the scientific terms for the common names.

    a. Shinbone _____

    b. Collarbone _____

    c. Hipbone _____

    d. Thighbone _____

After you have answered the questions for this section, you should be able to
- Classify joints according to their types.
- List the different types of joint movements.

15. Match the descriptions to these types of joints:

   fibrous joint   cartilaginous joint   synovial joint

   a. _____ lined with synovial membrane

   b. _____ immovable, as in a suture

   c. _____ slightly movable

   d. _____ freely movable

   e. _____ connected by hyaline cartilage

16. Fill in the table to describe the shoulder and elbow joints.

Joint	Anatomical Type	Degree of Movement
Shoulder joint	a.	b.
Elbow joint	c.	d

17. Using the alphabetized list of terms, label the following diagram to indicate the various types of joint movements.

   abduction, adduction, circumduction, eversion, extension, flexion, inversion, rotation

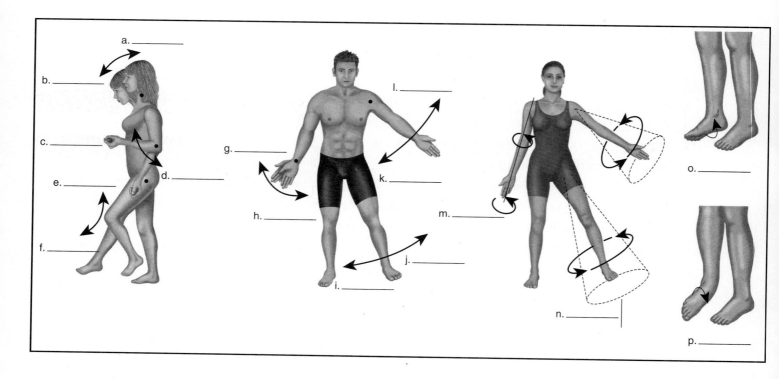

DEFINITIONS WORDSEARCH

Review key terms by completing this wordsearch using the following alphabetized list of terms:

```
O O S U F E M A E C A R T O P E R
E T J C O M P A C T A C I L B R Y
R O M M N A R I Y L A J O I N T N
C F U F T A R E T A L B L G N T E
N I E B A L O R S I I S S A S S A
U T T L N E T U S X U M G M O O B
O E S M E E T T S A L C O E T S O
A R O E L L L U Y O U A B N N R I
S T I O U T J S E S U S L T J U G
B A R E D B O N E M A R R O W M R
N A E M T A E R G C R R E F I N E
A P P E N D I C U L A R O V E A D
```

appendicular
axial
compact
fontanel
joint
ligament
osteoclast
periosteum
red bone marrow
suture

a. _____ Cell that breaks down old bone.

b. _____ Blood-forming tissue housed inside spongy bone.

c. _____ Portion of the skeleton made up of arms and legs.

d. _____ Portion of skeleton along midline.

e. _____ An articulation.

f. _____ Soft spot in newborn skull.

g. _____ Hardest portion of a long bone.

h. _____ Skull joint.

i. _____ Fibrous connective tissue covering each bone.

j. _____ Fibrous connective tissue band binding bone to bone.

CHAPTER TEST

OBJECTIVE TEST

Do not refer to the text when taking this test.

____ 1. The hardest portion(s) of a long bone is/are the
   a. articular cartilages.
   b. compact bone.
   c. spongy bone.
   d. ligaments.

____ 2. Bone-building cells are _____, while mature bone cells are _____.
   a. osteoblasts; osteocytes
   b. osteoclasts; osteoblasts
   c. osteoclasts; osteocytes
   d. osteoblasts; lacunae

____ 3. Where is the blood-forming tissue housed within a bone?
   a. throughout the bone
   b. along the periosteum
   c. within the compact bone
   d. within the spongy bone

____ 4. Which of these is a function of the skeleton?
   a. support
   b. protection
   c. production of red blood cells
   d. All of these are correct.

____ 5. Ligaments join
   a. bone to bone.
   b. muscle to muscle.
   c. muscle to bone.
   d. All of these are correct.

____ 6. Endochondral ossification involves
   a. long bones.
   b. a cartilaginous model.
   c. an epiphyseal plate.
   d. All of these are correct.

____ 7. Where in the body does intramembranous ossification occur?
  a. tibia
  b. humerus
  c. femur
  d. skull bones

____ 8. Choose the proper sequence describing repair of bone fractures.
  a. remodeling–fibrocartilage callus–hematoma–bony callus
  b. fibrocartilage callus–hematoma–remodeling–bony callus
  c. hematoma–fibrocartilage callus–bony callus–remodeling
  d. bony callus–remodeling–hematoma–fibrocartilage callus

____ 9. Which of the following hormones increases the amount of calcium in the blood?
  a. calcitonin
  b. vitamin D
  c. growth hormone
  d. parathyroid hormone

____ 10. Which of these is NOT a function of the skeleton?
  a. supports the body
  b. protects soft body parts
  c. stores minerals
  d. conducts nervous impulses

____ 11. Which of these is NOT in the appendicular skeleton?
  a. clavicle
  b. coxal bone
  c. metatarsals
  d. vertebrae

____ 12. Which of these is a facial bone?
  a. frontal bone
  b. occipital bone
  c. mandible
  d. All of these are correct.

____ 13. Mastoiditis, a complication of severe ear infections, is most apt to occur in which portion of the skull?
  a. fontanels
  b. frontal bone
  c. mastoid sinuses
  d. parietal bones

____ 14. Which bone articulates with all others of the cranium and is thus considered the keystone bone?
  a. ethmoid
  b. sphenoid
  c. temporal
  d. frontal

____ 15. Which of these is NOT a bone in the lower limb?
  a. femur
  b. tibia
  c. ulna
  d. fibula

____ 16. Vertebrae have
  a. immovable joints.
  b. freely movable joints.
  c. slightly movable joints.
  d. joints that vary from one person to the next.

____ 17. The shoulder joint is an example of a _____joint.
  a. synovial
  b. freely movable
  c. ball-and-socket
  d. All of these are correct.

____ 18. The bones making up your palm are the
  a. phalanges.
  b. carpals.
  c. metacarpals.
  d. metatarsals.

____ 19. Which vertebrae support the lower back?
  a. cervical
  b. lumbar
  c. thoracic
  d. cranial

____ 20. Which type of joint is most likely to develop arthritis?
  a. fibrous
  b. cartilaginous
  c. synovial
  d. epiphyseal disk

____ 21. Fluid-filled sacs within joints are called
  a. bursae.
  b. menisci.
  c. articulations.
  d. synovial fluid.

____ 22. Bending your knee so that your heel touched your buttock would be called
  a. flexion.
  b. extension.
  c. eversion.
  d. pronation.

____ 23. The system that provides calcium and other nutrients for bone growth and repair of the skeleton is the _____ system.
  a. urinary
  b. digestive
  c. endocrine
  d. lymphatic

____ 24. The system that has receptors to send information to and from bones and joints is the _____ system.
  a. respiratory
  b. integumentary
  c. nervous
  d. cardiovascular

Use the space provided to answer these questions in complete sentences.

25. Why are knee injuries usually serious?

26. Describe the remodeling of bone.

Test Results: _____ number correct ÷ 26 = _____ × 100 = _____%

ANSWER KEY

STUDY QUESTIONS

1. a. lamellae **b.** spongy bone **c.** red bone marrow **d.** fibrous connective **2. a.** hyaline cartilage; ends of bones **b.** fibrocartilage; intervertebral disks **c.** elastic cartilage; ear flaps **3. a.** hyaline cartilage **b.** compact bone **c.** spongy bone **d.** medullary cavity **4. a.** spongy bone **b.** compact bone **c.** hyaline cartilage **d.** medullary cavity **5. a.** bone-forming osteoblasts eventually become mature osteocytes **b.** to break down bone **c.** ossification **d.** They allow a bone to grow longer. **6. a.** I **b.** E **c.** E **d.** I **e.** I **7. a.** calcium **b.** exercise **c.** hormones **8.** d, a, b, c **9. a.** supports the body **b.** protects soft body parts **c.** produces blood cells **d.** stores minerals and fats **e.** permits body movement **10. a.** ap, pel **b.** ax **c.** ap **d.** ap, pec **e.** ax **f.** ap **g.** ax **h.** ap **i.** ap, pec **j.** ap **k.** ap **l.** ap **11. a.** frontal **b.** occipital **c.** temporal **d.** mandible **e.** maxillae **f.** zygomatic **12. a.** to protect the spinal cord **b.** thoracic **c.** lumbar **d.** cervical **13. a.** tibia **b.** fibula **c.** humerus **d.** metatarsals **14. a.** tibia **b.** clavicle **c.** coxal bone **d.** femur **15. a.** synovial **b.** fibrous **c.** cartilaginous **d.** synovial **e.** cartilaginous **16. a.** ball-and-socket (synovial) **b.** freely movable **c.** hinge (synovial) **d.** freely movable **17. a.** extension **b.** flexion **c.** flexion **d.** extension **e.** flexion **f.** extension **g.** abduction **h.** adduction **i.** adduction **j.** abduction **k.** adduction **l.** abduction **m.** rotation **n.** circumduction **o.** inversion **p.** eversion

DEFINITIONS WORDSEARCH

```
        F
    C O M P A C T           L
M   N           L   J O I N T
U   T       E       A       G
E   A       R   I           A
T   N       U   X           M
S   E       T S A L C O E T S O
O   L       U               N
I   S                       T
R E D B O N E M A R R O W
E
A P P E N D I C U L A R
```

a. osteoclast **b.** red bone marrow **c.** appendicular **d.** axial **e.** joint **f.** fontanel **g.** compact **h.** suture **i.** periosteum **j.** ligament

CHAPTER TEST

1. b **2.** a **3.** d **4.** d **5.** a **6.** d **7.** d **8.** c **9.** d **10.** d **11.** d **12.** c **13.** c **14.** b **15.** c **16.** c **17.** d **18.** c **19.** b **20.** c **21.** a **22.** a **23.** b **24.** c **25.** The knee is a complex, large joint with many ligaments and menisci for strength and stability. Injuring just one area of the knee means much pain and time spent waiting for healing. The knee also supports the weight of the body, which can make healing a very slow process. **26.** Osteoclasts tear down worn-out bone tissue. During this process, calcium is released into the bloodstream. Osteoblasts build and remodel the bone according to the needs of the body.

11

MUSCULAR SYSTEM

The muscular system performs several functions for your body aside from movement. This chapter focuses on the anatomy and physiology of skeletal muscle.

Review the functions of skeletal muscles (p. 198). Understand that one end of the muscle is called the insertion, while the other is the origin (p. 199). Muscles work in groups, some together and others in opposition (p. 199). Spend time learning about how muscles are named (p. 200), which will help you remember names of the major muscles of the body (p. 201) and their function (p. 201). Cover up muscle names in Figure 11.3 and quiz yourself.

Within skeletal and cardiac muscle cells there are contractile units called sarcomeres. Study their structure, then learn how contraction occurs as filaments within sarcomeres slide past each other (p. 202). You should be able to describe a neuromuscular junction and how it functions (p. 204).

Briefly review your notes on cellular respiration from Chapter 3 before you study the section on energy for muscle contraction (p. 208). Recall the function of ATP in the cell. ATP is needed by muscle fibers for contraction. Muscle fibers have several ways to acquire ATP, as you will discover as you read Section 11.4 in the text.

The chapter includes a Health Focus on the benefits of exercise.

Study the *Human Systems Work Together* illustration on page 213 and make a list of ways the muscular system helps maintain homeostasis. Also, determine how the other systems assist the muscular system in maintaining homeostasis.

TIP: Be sure to visit the Online Learning Center that accompanies *Human Biology* 9/e. It has practice quizzes, interactive activities, labeling exercises, art quizzes, animations, flash cards, and much more. http://www.mhhe.com/maderhuman9

STUDY QUESTIONS

Study the text section by section. Answer the study questions so that you can fulfill the learning objectives for each section.

11.1 TYPES AND FUNCTIONS OF MUSCLES (PAGES 198–201)

After you have answered the questions for this section, you should be able to
- Describe the three types of muscle tissue.
- List the functions of skeletal muscle.
- Name the major muscles of the body, and know their actions.

1. Match the phrases to the three types of muscle tissue:
   smooth   cardiac   skeletal

   a. _____ constant, rhythmical contraction

   b. _____ contraction under voluntary control

   c. _____ uninucleated, spindle-shaped cells

   d. _____ found in walls of internal organs

   e. _____ branched and joined at intercalated disks

2. List three functions of skeletal muscle.

   a._____

   b._____

   c._____

3. Individual muscles are wrapped in a coat of fibrous connective tissue, called a. _____. When a group of muscles work together, the one pulling the greatest load is called the b. _____, while the other helper muscles are called c. _____. Muscles that work opposite to each other are known as d. _____.

4. Using the alphabetized list of terms, label the following diagram of muscles and bones in the upper limb.

biceps brachii
humerus
insertion
origin
radius
triceps brachii
ulna

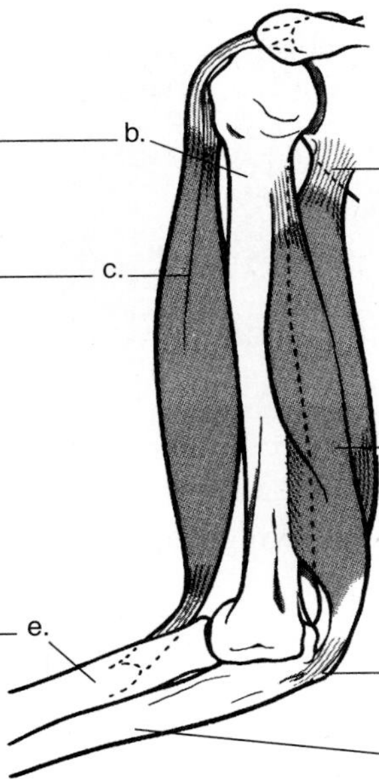

b. _____

a. _____

c. _____

d. _____

e. _____

f. _____

g. _____

Use the space provided to answer this question using a complete sentence.

5. With reference to the muscles in the drawing for question 4, why are the biceps brachii and triceps brachii antagonistic pairs? _____

_____

6. Name the thigh muscles that act as antagonists.

a. Front of thigh_____

b. Back of thigh_____

7. Name the antagonistic muscles of the lower limb.

a. Front of lower limb_____

b. Back of lower limb_____

11.2 MECHANISM OF SKELETAL MUSCLE FIBER CONTRACTION (PAGES 202–205)

After you have answered the questions for this section, you should be able to
- Describe the anatomy of a muscle fiber.
- Explain how the sarcomere shortens during muscle contraction.
- Describe the neuromuscular junction, and tell how impulses are transferred to the sarcolemma.

8. The plasma membrane of a muscle fiber is named the a._____. A special kind of endoplasmic reticulum, called b._____ _____, stores the element c. _____ in the cell. d. _____ are invaginations of the plasma membrane.

9. Muscle cells, known as a. _____, contain thin, or b. _____, filaments and thick, or c. _____, filaments. The myofilaments are arranged in contractile units called d. _____.

10. Using the alphabetized list of terms, label the following diagram of a portion of a muscle fiber.

myofibril
sarcolemma of muscle fiber
sarcomere
sarcoplasmic
    reticulum
T tubule
Z line
    (used twice)

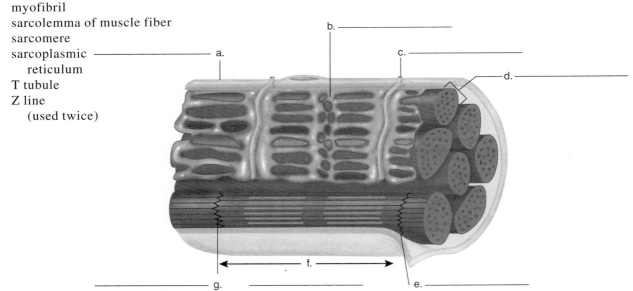

a. _____
b. _____
c. _____
d. _____
e. _____
f. _____
g. _____

11. Using the alphabetized list of terms, label the following diagram of a sarcomere.

actin filament
H zone
myosin filament
Z line

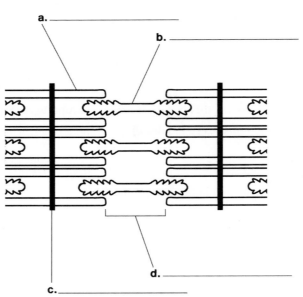

a. _____
b. _____
c. _____
d. _____

12. Which of your labels in question 11 is a thin filament? a. _____ Which of your labels is a thick filament? b. _____ Which of your labels is reduced in size when a sarcomere contracts? c. _____ Which component has cross-bridges? d. _____ Which of your labels is the filament that moves when the sarcomere contracts? e. _____ What molecule immediately supplies energy for muscle contraction? f. _____

13. What is the proper sequence for these phrases to describe what occurs at the neuromuscular junction to trigger muscle contraction? Indicate by letter. _____
    a. receptor sites on sarcolemma
    b. nerve impulse
    c. release of calcium from sarcoplasmic reticulum
    d. neurotransmitter acetylcholine released
    e. sarcomeres shorten
    f. synaptic cleft
    g. spread of impulses over sarcolemma to T tubules

14. Study the following diagram and fill in the blanks that follow.

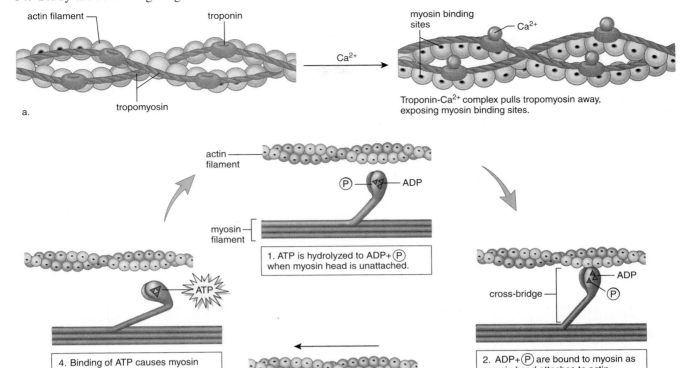

a.

b.

When calcium ions are released they bind to a. _____, causing threads of b. _____ to shift and expose c. _____ binding sites. Myosin heads split ATP, and then they bind to d. _____, forming a e. _____. The release of ADP + ℗ from the myosin heads causes the cross-bridges to f. _____ their position and g. _____ the actin filaments toward the middle of the sarcomere.

11.3 WHOLE MUSCLE CONTRACTION (PAGE 206)

After you have answered the questions for this section, you should be able to
• Describe the basic laboratory experiments on whole muscle contraction.
• Explain how muscle tone is maintained.

15. Using the alphabetized list of terms, label the following diagram of a single muscle twitch.
    contraction
    latent period
    relaxation

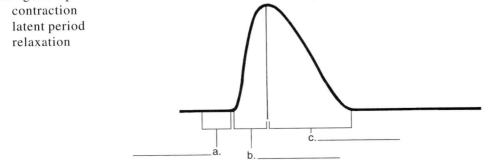

a. _____ b. _____ c. _____

16. On the following myogram, indicate the locations where the stimulus was applied, where fatigue begins, where tetanus occurs, and where the time interval is shown.

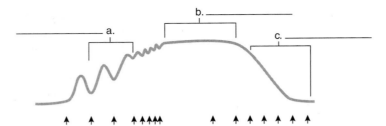

17. How do you recognize fatigue of a muscle? _____

18. Stimulation of a muscle fiber results in a contraction that can be described as _____.

19. Describe these terms and state their significance.

 a. tone _____

 _____

 b. motor unit _____

 _____

11.4 ENERGY FOR MUSCLE CONTRACTION (PAGES 208–209)

After you have answered the questions for this section, you should be able to
- List the sources of energy available within a muscle fiber, and explain the circumstances in which each is used.
- Discuss the events leading to oxygen debt.

20. Compare sources of energy and oxygen for muscle contraction. Match the statements to these terms:

 ATP   creatine phosphate   myoglobin   fermentation

 a. _____ anaerobic source of energy

 b. _____ produced during cellular respiration

 c. _____ used to regenerate ATP from ADP

 d. _____ stores oxygen in muscle fibers

 e. _____ results in the buildup of lactate and results in an oxygen deficit

21. Oxygen deficit occurs when a. _____ is used up in muscles, and the blood does not

 supply b. _____ rapidly enough. Why are marathon runners less apt to become exhausted

 due to an oxygen deficit? c. _____

11.5 MUSCULAR DISORDERS (PAGE 210)

After you have answered the questions for this section, you should be able to
- Describe common muscular injuries and diseases.

22. Match the description to the types of muscular disorder:
 spasm   sprain   strain   muscular dystrophy   myalgia   myasthenia gravis   tendinitis
 a. _____ inflamed tendon, painful joint movement

 b. _____ progressive weakening of muscles

 c. _____ overstretching of a muscle near a joint

 d. _____ achy muscles

 e. _____ sudden, involuntary muscular contraction

 f. _____ autoimmune disease

 g. _____ injury of muscle, ligament, tendon, blood vessels, and nerve due to twisting of joint

After you have answered the questions for this section, you should be able to
- Discuss how the muscular system works with other systems of the body to maintain homeostasis.

23. The *Human Systems Work Together* diagram in your textbook shows how the muscular system works with the other organ systems of the body. Match the organ systems with the correct descriptions.
  - (1) Muscle contraction keeps blood moving in heart and blood vessels.
  - (2) Muscle contraction causes bones to move joints; muscles help protect bones.
  - (3) Muscle contraction provides heat to warm skin.
  - (4) Skeletal muscle contraction moves lymph; physical exercise enhances immunity.
  - (5) Muscles help protect glands.
  - (6) Smooth muscle contraction accounts for peristalsis; skeletal muscles support and help protect abdominal organs.
  - (7) Muscle contraction assists breathing; physical exercise increases respiratory capacity.
  - (8) Muscle contraction occurs during orgasm and moves gametes; abdominal and uterine muscle contraction occurs during childbirth.
  - (9) Smooth muscle contraction assists voiding of urine; skeletal muscles support and help protect urinary organs.
  - (10) Muscle contraction moves eyes, permits speech, and creates facial expressions.

  a. _____ integumentary system
  b. _____ lymphatic system/immunity
  c. _____ cardiovascular system
  d. _____ skeletal system
  e. _____ urinary system
  f. _____ nervous system
  g. _____ respiratory system
  h. _____ reproductive system
  i. _____ endocrine system
  j. _____ digestive system

DEFINITIONS WORDSEARCH

Review key terms by completing this wordsearch using the following alphabetized list of terms:

```
T E L L E Y I N I T C A
E D N S O R R Y T I Y M
T A O I A M H I E S O M
A T D Y B P A B T O N E
N H N M O R I G I N P L
U E E R O L M E T O L O
S M T E V I H T I W O C
S A R C O M E R E P P R
B M Y O S I N S T E I A
Y N O I T R E S N I N S
```

actin
insertion
tone
myosin
origin
sarcolemma
sarcomere
tendon
tetanus
tone

  a. _____ Structural and functional unit of a myofibril.
  b. _____ End of muscle attached to the movable bone.
  c. _____ End of muscle attached to immovable bone.
  d. _____ Plasma membrane of a muscle cell.
  e. _____ Tension that is still present in resting muscle.
  f. _____ Protein of thin filaments.
  g. _____ Protein of thick filaments.
  h. _____ Joins muscle to bone.
  i. _____ Sustained maximal muscle contraction.

Do not refer to the text when taking this test.

_____ 1. The fascia surrounding a muscle is made up of
   a. cartilage.
   b. fibrous connective tissue.
   c. blood vessels.
   d. adipose tissue.

_____ 2. The contractile unit of a myofibril is the
   a. neuromuscular junction.
   b. actin filament.
   c. sarcomere.
   d. myosin filament.

_____ 3. A muscle pulling just opposite to another can be best described as a(n)
   a. protagonist.
   b. antagonist.
   c. prime mover.
   d. synergist.

_____ 4. Which of these is a function of skeletal muscle?
   a. heat production
   b. posture
   c. movement
   d. All of these are correct.

_____ 5. Choose the muscle that is the antagonist to the triceps brachii.
   a. gastrocnemius
   b. quadriceps femoris
   c. biceps brachii
   d. latissimus dorsi

_____ 6. Choose the muscle that is the antagonist to the tibialis anterior.
   a. gastrocnemius
   b. sartorius
   c. gluteus maximus
   d. trapezius

_____ 7. Which muscle covers the shoulder and functions to raise the arm?
   a. external oblique
   b. deltoid
   c. occipitalis
   d. peroneus longus

_____ 8. Which muscle is located on the front thigh and raises (extends) the lower limb?
   a. trapezius
   b. rectus abdominis
   c. pectoralis major
   d. quadriceps femoris

_____ 9. Which portion of the muscle is on the stationary bone?
   a. the insertion
   b. the origin
   c. the bursa
   d. the belly

_____ 10. Which of these is the smallest unit?
   a. muscle fiber
   b. myofibril
   c. sarcomere
   d. actin

_____ 11. When sarcomeres contract, they get shorter, and this requires that muscles work in antagonistic pairs.
   a. true
   b. false

_____ 12. Myosin is
   a. the thick filament of a sarcomere.
   b. a protein.
   c. an organelle.
   d. Both _a_ and _b_ are correct.

_____ 13. Which of these contains the cross-bridges?
   a. actin
   b. myosin
   c. Z line
   d. All of these are correct.

_____ 14. According to the sliding filament theory,
   a. actin moves past myosin.
   b. myosin moves past actin.
   c. both myosin and actin move past each other.
   d. None of these is correct.

_____ 15. The release of calcium from the sarcoplasmic reticulum
   a. causes sarcomeres to relax.
   b. causes sarcomeres to contract.
   c. is the result of a nervous impulse to contract.
   d. Both _b_ and _c_ are correct.
   e. All of these are correct.

_____ 16. When a nervous impulse reaches the axon terminal,
   a. neurotransmitters are released into the synaptic cleft.
   b. calcium is released from the sarcoplasmic reticulum.
   c. calcium is stored in the T tubules.
   d. neurotransmitter is released by the sarcolemma.

_____ 17. The junction between nerve and muscle is
   a. called the neuromuscular junction.
   b. where a neurotransmitter travels from the nerve fiber to the muscle fiber.
   c. the region where nerves innervate muscles.
   d. All of these are correct.

_____ 18. The Z line of a sarcomere is where the
   a. cross-bridges attach.
   b. actin filaments attach.
   c. myosin filaments attach.
   d. nerve innervates the muscle cell.

____19. Muscle fatigue
   a. follows summation and tetanus.
   b. involves the buildup of lactate.
   c. occurs only in the laboratory.
   d. Both *a* and *b* are correct.
   e. All of these are correct.
____20. Oxygen deficit may be associated with
   a. anaerobic cellular respiration.
   b. fermentation.
   c. cramping.
   d. lactate metabolism.
   e. All of these are correct.
____21. Creatine phosphate is
   a. used by sarcomeres.
   b. used to change ADP to ATP.
   c. a molecule found in DNA.
   d. All of these are correct.

____22. Which system provides oxygen and removes carbon dioxide from muscles?
   a. integumentary
   b. skeletal
   c. respiratory
   d. urinary
____23. The muscular system aids the flow of materials through vessels of which system?
   a. lymphatic
   b. cardiovascular
   c. urinary
   d. All of these are correct.

THOUGHT QUESTIONS

Use the space provided to answer these questions in complete sentences.

24. Why does muscle contraction depend on the presence of myofibrils in a muscle fiber?

25. What events lead to sarcomere contraction after ACh is released at a neuromuscular junction?

Test Results: _____ number correct ÷ 25 = _____ × 100 = _____ %

ANSWER KEY

STUDY QUESTIONS

1. a. cardiac **b.** skeletal **c.** smooth **d.** smooth **e.** cardiac
2. a. provides posture **b.** movement **c.** provides heat
3. a. fascia **b.** prime mover **c.** synergists **d.** antagonists
4. a. origin **b.** humerus **c.** biceps brachii **d.** triceps brachii
e. radius **f.** insertion **g.** ulna   **5.** The biceps brachii raises
and the triceps brachii lowers the forearm.   **6. a.** quadriceps femoris group **b.** hamstring muscles   **7. a.** tibialis
anterior **b.** gastrocnemius   **8. a.** sarcolemma **b.** sarcoplasmic reticulum **c.** calcium **d.** T tubules   **9. a.** fibers
b. actin **c.** myosin **d.** sarcomeres   **10. a.** sarcolemma of
muscle fiber **b.** sarcoplasmic reticulum **c.** T tubule
d. myofibril **e.** Z line **f.** sarcomere **g.** Z line   **11. a.** actin
filament **b.** myosin filament **c.** Z line **d.** H zone
12. a. actin filament **b.** myosin filament **c.** H zone
d. myosin **e.** actin filament **f.** ATP   **13.** b, d, f, a, g, c, e
14. a. troponin **b.** tropomyosin **c.** myosin **d.** actin
e. cross-bridge **f.** change **g.** pull   **15. a.** latent period
b. contraction **c.** relaxation   **16. a.** stimulus applied **b.**
tetanus occurs **c.** fatigue begins **d.** time interval   **17.** The
muscle relaxes even though a stimulus has been ap-
plied.   **18.** all-or-none   **19. a.** Some muscle fibers in
the body are always contracting; keeps body from col-
lapsing. **b.** A motor neuron stimulates several fibers at once;
allows degree of contraction and the fewer fibers in motor
units, the finer the control of that body part.   **20. a.** fermentation, creatine phosphate **b.** ATP **c.** creatine phosphate **d.** myoglobin **e.** fermentation   **21. a.** ATP
b. oxygen **c.** Marathon runners have more mitochondria.
22. a. tendinitis **b.** muscular dystrophy **c.** strain **d.** myalgia **e.** spasm **f.** myasthenia gravis **g.** sprain   **23. a.** (3)
b. (4) **c.** (1) **d.** (2) **e.** (9) **f.** (10) **g.** (7) **h.** (8) **i.** (5) **j.** (6)

DEFINITIONS WORDSEARCH

```
T           N I T C A
E    N                M
T    O                M
A    D        T O N E
N    N  O R I G I N   L
U    E                O
S    T                C
S A R C O M E R E     R
  M Y O S I N         A
  N O I T R E S N I   S
```

a. sarcomere   **b.** insertion   **c.** origin   **d.** sarcolemma
e. tone   **f.** actin   **g.** myosin   **h.** tendon   **i.** tetanus

CHAPTER TEST

1. b   **2.** c   **3.** b   **4.** d   **5.** c   **6.** a   **7.** b   **8.** d   **9.** b
10. d   **11.** a   **12.** d   **13.** b   **14.** a   **15.** d   **16.** a
17. d   **18.** b   **19.** d   **20.** e   **21.** b   **22.** c   **23.** d
24. Myofibrils contain the contractile proteins, namely
myosin and actin filaments.   **25.** Impulses travel down
the T system of a muscle fiber. Ca^{2+} is released from the
sarcoplasmic reticulum. Myosin cross-bridges attach to
and pull the actin filaments past myosin filaments in each
sarcomere.

PART 4 INTEGRATION AND COORDINATION IN HUMANS

12

NERVOUS SYSTEM

This is a challenging chapter that will require diligent study on your part. Begin by using a motor neuron to learn the three parts of all neurons. Use Figure 12.2 to see how the three types of neurons relate to one another.

Be able to associate the movement of ions across the axomembrane with the graph of the nerve impulse (action potential) depicted in Figure 12.4c. Transmission of the nerve impulse between neurons is clearly explained in Figure 12.5. Study this figure to gain an understanding of synapse structure and function.

Figure 12.7 depicts the organization of the nervous system. Be sure you can explain the positioning of the arrows in this figure. The central nervous system includes the spinal cord and brain. Page 225 gives you a good overview of the spinal cord. Use Figure 12.9 to learn the parts of the brain, and make a chart that lists the parts and their functions. You also want to have a thorough understanding of Figure 12.16, which depicts a reflex arc. Be able to contrast the somatic system with the autonomic system and then use Table 12.2 to compare the somatic system with the two divisions of the autonomic system.

The nervous system, together with the endocrine system, coordinates the functioning of the other systems in the body.

TIP: Be sure to visit the Online Learning Center that accompanies *Human Biology* 9/e. It has practice quizzes, interactive activities, labeling exercises, art quizzes, animations, flash cards, and much more. http://www.mhhe.com/maderhuman9

STUDY QUESTIONS

Study the text section by section. Answer the study questions so that you can fulfill the learning objectives for each section.

12.1 NERVOUS TISSUE (PAGES 218–223)

After you have answered the questions for this section, you should be able to
- Describe the structure and function of the three major types of neurons.
- Describe the formation of an action potential.
- Describe what occurs at a synapse.

1. Every neuron has these three parts. What is the function of each?

   a. dendrite _____

   b. cell body _____

   c. axon _____

2. In the following diagram, label the parts of the sensory neuron, the interneuron, and the motor neuron, using the following alphabetized list of terms. *Note:* Some terms may be used more than once.

axon   axon terminal   cell body   dendrites   effector   node of Ranvier
nucleus of Schwann cell   sensory receptor

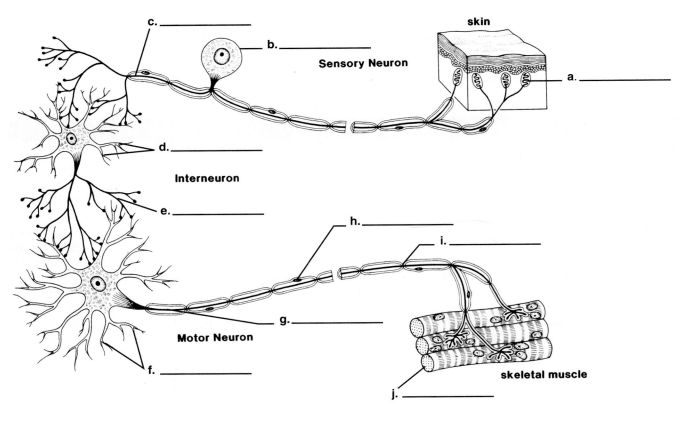

3. State the function of the complete sensory neuron. _____
_____

4. State the function of the complete interneuron. _____
_____

5. State the function of the complete motor neuron. _____
_____

6. Examine this diagram depicting the resting potential, and then answer the questions that follow:

a. What is an oscilloscope? _____

_____

b. The oscilloscope is reading −65 mV. Why is the inside of an axon negative compared to the outside?

_____

c. At the time of the resting potential, are there more or less Na⁺ ions outside the axon than inside the axon?

_____

d. At the time of the resting potential, are there more or less K⁺ ions outside the axon than inside the axon?

_____

e. What mechanism accounts for the unequal distribution of ions across the axomembrane?

_____

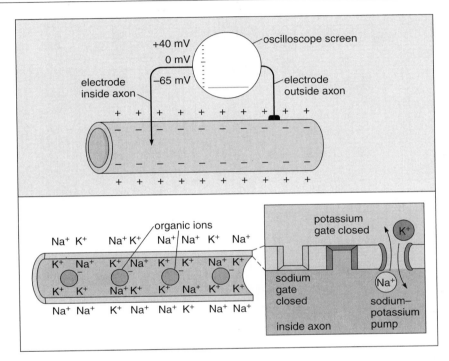

7. Label the following diagram depicting the trace on the oscilloscope screen as an action potential occurs using the alphabetized list of terms.

action potential
depolarization
repolarization
resting potential
threshold

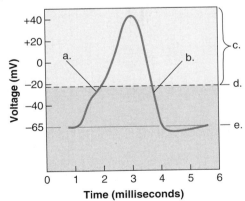

8. During the time of rest, the _____-_____ pump restores the original distribution of ions across the membrane of a nerve fiber.

9. Label the following diagrams, using these terms:

   axon terminals   dendrite   neurotransmitter   synaptic cleft   synaptic vesicles

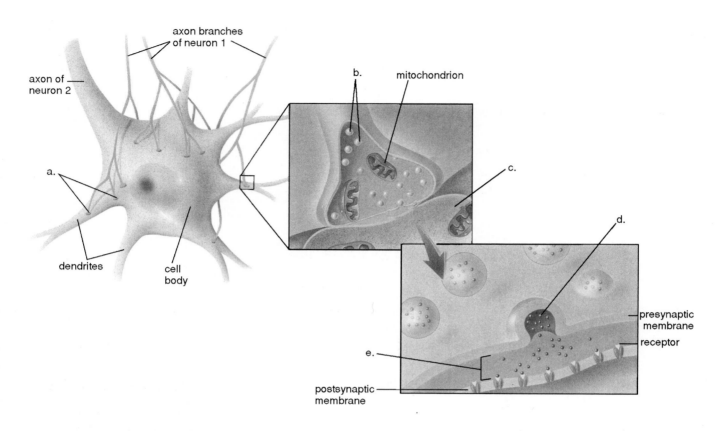

10. What causes the transmission of the nerve impulse across a synapse? _____

    _____

11. Indicate whether these statements are true (T) or false (F).
    a. _____ A single neuron synapses with one other neuron.
    b. _____ A single neuron synapses with many other neurons.
    c. _____ Excitatory signals have a hyperpolarizing effect, and inhibitory signals have a depolarizing effect.
    d. _____ Integration is the summing up of excitatory and inhibitory signals.
    e. _____ The more inhibitory signals received, the more likely the axon will conduct a nerve impulse.

After you have answered the questions for this section, you should be able to
- Label a cross section of the spinal cord, and give two functions of the spinal cord.
- Describe, in general, the anatomy of the brain, name five major parts, and give a function of each.
- Name the lobes of the cerebrum, and give a function of each.

12. Label the parts of the spinal cord using the alphabetized list of terms.

central canal
dorsal root
gray matter
ventral root
white matter

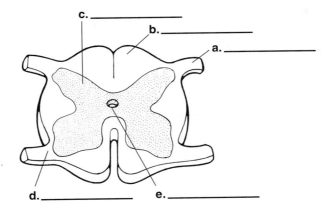

c. _____
b. _____
a. _____
d. _____
e. _____

13. Indicate whether the following statements are true (T) or false (F). Rewrite all false statements as true statements.

a. _____ White matter is white because it contains cell bodies of interneurons that run together in bundles called tracts. Rewrite: _____

b. _____ The spinal cord carries out the integration of incoming information before sending signals to other parts of the nervous system. Rewrite: _____

c. _____ When the spinal cord is severed, we suffer a loss of sensation but not a loss of voluntary control. Rewrite: _____ _____

14. Label the parts of the brain, using the following alphabetized list of terms.

cerebellum   cerebrum   corpus callosum   medulla oblongata   pituitary gland   pons   thalamus

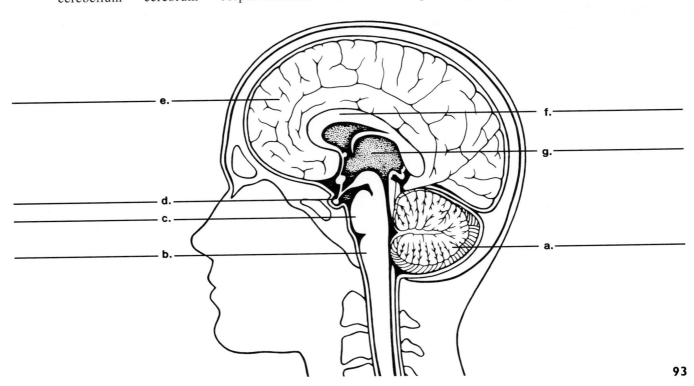

e. _____
f. _____
g. _____
d. _____
c. _____
a. _____
b. _____

15. Fill in the following table to indicate the functions of the parts of the brain.

Brain Part	Function
Cerebrum	a.
Thalamus	b.
Hypothalamus	c.
Cerebellum	d.
Medulla oblongata	e.

16. Match each description to the correct name of the lobe, using these terms:

      all lobes   frontal lobe   occipital lobe   parietal lobe   temporal lobe

Note: Some terms may be used more than once.

a. _____ Contains primary motor area, which controls voluntary motions.

b. _____ Contains primary somatosensory area, which receives sensory information from the skin and skeletal muscles.

c. _____ Contains a primary visual area.

d. _____ Contains a primary auditory area.

e. _____ Contains a primary association area.

f. _____ Carries on higher mental functions such as reasoning and critical thinking.

12.3 THE LIMBIC SYSTEM AND HIGHER MENTAL FUNCTIONS (PAGES 230–231)

After you have answered the questions for this section, you should be able to
- Describe in general the structure and the function of the limbic system.
- List and describe the different types of memory.
- Tell how the limbic system assists memory storage and retrieval.
- Describe our present understanding of the human ability to use language and speak.

17. Place a check beside those structures that are a part of the limbic system.
 a. _____ tracts that join portions of the cerebral lobes, subcortical nuclei, and diencephalon
 b. _____ hippocampus, which functions in retrieving memories
 c. _____ amygdala, which adds emotional overtones to memories

18. Which of these best describes the limbic system? _____
 a. a system that involves reasoning
 b. a system that involves emotions
 c. a system that involves memories
 d. All of these are correct.

19. Study the following diagram and answer the questions.

prefrontal area ↔ hippocampus and amygdala ↔ cortical sensory areas ← Sensory inputs

hippocampus and amygdala ↕ diencephalon

The hippocampus and amygdala are in contact with what areas of the brain?

a. _____, b. _____, c. _____

In which of these are memories stored for later retrieval? d. _____

In which of these areas are memories used to plan future actions? e. _____

In 19f–h, use the space provided to answer these questions using a complete sentence.

What is the role of the hippocampus? f. _____
_____

What is the role of the amygdala? g. _____
_____

Why are there two sets of arrows in the diagram? h. _____
_____

20. Indicate whether these statements are true (T) or false (F).
   a. _____ Broca's area is a motor speech area.
   b. _____ Damage to Wernicke's area results in the inability to comprehend speech.
   c. _____ Only the left side of the brain contains a Broca's area and a Wernicke's area.

12.4 THE PERIPHERAL NERVOUS SYSTEM (PAGES 232–235)

After you have answered the questions for this section, you should be able to
• Define a nerve, and distinguish between spinal and cranial nerves.
• Describe the path of a reflex arc in the somatic system.
• Contrast the somatic system with the autonomic system.
• Describe the autonomic system, and cite similarities as well as differences in the structure and function of the two divisions.

21. Name two types of nerves in the peripheral nervous system: a. _____ and b. _____ nerves. What is

a nerve? c. _____ Why is a spinal nerve called a

mixed nerve? d. _____

22. Label this diagram of the reflex arc, using the following alphabetized list of terms.
   dorsal-root ganglion
   effector
   interneuron
   motor neuron
   sensory neuron
   sensory receptor

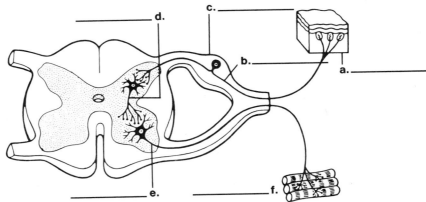

95

23. A stimulus is received by a. _____ in the skin, and they generate nerve impulses in the fibers of b. _____. Next the impulses are picked up by c. _____ in the spinal cord. Then they are received by the fibers of d. _____, which stimulate a muscle to contract, causing a reaction to the stimulus.

24. Explain how the brain becomes aware of automatic reflex actions. _____
_____

25. Indicate three ways in which the sympathetic and parasympathetic systems are similar.

a. _____

b. _____

c. _____

26. Indicate ways in which the sympathetic and parasympathetic systems differ by filling in the following table.

	Sympathetic	Parasympathetic
Type of situation	a.	b.
Neurotransmitter	c.	d.
Ganglia near cord or ganglia near organ?	e.	f.
Spinal nerves only or spinal nerves plus vagus?	g.	h.

12.5 DRUG ABUSE (PAGES 237–239)

After you have answered the questions for this section, you should be able to
• Describe drug action in general, and discuss the effects of alcohol, marijuana, cocaine, and heroin.

27. Drugs are believed to affect, in particular, what part of the brain? a. _____ There are both inhibitory and excitatory neurotransmitters in the brain. If a drug blocks the action of an inhibitory neurotransmitter, what psychological effect will it have? b. _____ If a drug blocks the action of an excitatory neurotransmitter, what psychological effect will it have? c. _____

28. The drug a. _____ in tobacco products causes neurons to release dopamine, which reinforces dependence on this drug. Occasional b. _____ users experience euphoria, alterations in vision and judgment, and distortions of space and time. Heavy use of c. _____ often leads to liver damage. d. _____ is a ready-to-smoke form of cocaine.

29. According to endorphin research, what causes heroin withdrawal symptoms? _____
_____

For each correct answer, Simon says, "You may move one step forward." Total possible number of steps forward is 10 steps.

1. Which of these would NOT be used when studying nerve conduction?
   a. voltmeter
   b. oscilloscope
   c. electron microscope
   d. electrodes
   e. electric current

2. Which one is NOT directly needed for nerve conduction?
   a. dendrites
   b. axons
   c. axomembrane
   d. nucleus
   e. axoplasm
   f. ions

3. Which one does NOT move during nerve conduction?
   a. sodium
   b. potassium
   c. positive charges
   d. negative charges

4. Which one does NOT accurately describe a resting neuron?
   a. positive on the outside of the membrane and negative on the inside
   b. Na^+ on the outside of the membrane and K^+ on the inside
   c. −65 mV inside
   d. negative on both sides of the membrane

5. Which one is NOT involved with an action potential?
   a. resting potential
   b. permeability
   c. sodium-potassium pump
   d. axomembrane
   e. acetylcholine
   f. ions
   g. glycogen

6. Which one does NOT conduct a nerve impulse?
   a. sensory neurons
   b. osteocytes
   c. motor neurons
   d. sensory nerves
   e. motor nerves

7. Which pair is improperly matched?

8. Which number could NOT be associated with an action potential?
   a. −65 mV
   b. 0 mV
   c. +40 mV
   d. −40 watts

9. Which pair is improperly matched?
   a. e^-—nerve impulse
   b. sodium-potassium pump—resting potential
   c. +—Na^+
   d. −—K^+
   e. plasma membrane—semipermeable

10. Which one is NOT true?

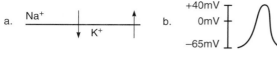

a. Na⁺ → K⁺

b. +40mV / 0mV / −65mV

c. + + + + + + / − − − − − −

d. Mg⁺ Mg⁺ Mg⁺ / Cl⁻ Cl⁻ Cl⁻

How many steps were you allowed by Simon? _____

DEFINITIONS WORDSEARCH

Review key terms by completing this wordsearch using the following alphabetized list of terms:

```
G A N G L I O N P O L I C
C D E T H A L A M U S E C
N G N L L I M B I C Y Y I
S E O K I M D F C H N H T
C E R E B E L L U M A T A
H T U X E L F E R L P N M
U D E N D R I T E O S K O
N K N G O D L O V E E I S
A C E T Y L C H O L I N E
```

acetylcholine
cerebellum
dendrite
ganglion
limbic
neuron
reflex
somatic
synapse
thalamus

a. _____ Portion of brain in third ventricle; integrates sensory input.

b. _____ Nerve cell.

c. _____ Dorsally located, a part of the brain that coordinates graceful skeletal mus–cle movement.

d. _____ Extension of a neuron that conducts signals toward the cell body.

e. _____ Automatic, involuntary response to a stimulus.

f. _____ Region between two neurons where impulses pass.

g. _____ System in brain concerned with emotions and memory.

h. _____ A neurotransmitter.

i. _____ Portion of peripheral nervous system controlling skeletal muscles.

j. _____ Collection of cell bodies within the peripheral nervous system.

CHAPTER TEST

OBJECTIVE TEST

Do not refer to the text when taking this test.

____ 1. Sensory neurons
    a. take impulses to the CNS.
    b. take impulses away from the CNS.
    c. have a cell body in the dorsal root ganglion.
    d. Both *a* and *c* are correct.

____ 2. Which of the following neurons would be found in the autonomic system of the PNS?
    a. motor neurons ending at skeletal muscle
    b. motor neurons surrounding the esophagus
    c. sensory neurons at the surface of the skin
    d. sensory neurons attached to olfactory receptors
    e. interneurons in the spinal cord

____ 3. The neuron that is found wholly and completely within the CNS is the
    a. motor neuron.
    b. sensory neuron.
    c. interneuron.
    d. All of these are correct.

____ 4. Which of these contains the nucleus?
    a. axon
    b. dendrite
    c. cell body
    d. Any of these may contain the nucleus.

_____ 5. The downswing of the nervous impulse is caused by the movement of
   a. sodium ions to the inside of a neuron.
   b. sodium ions to the outside of a neuron.
   c. potassium ions to the inside of a neuron.
   d. potassium ions to the outside of a neuron.

_____ 6. The resting potential is maintained by the sodium-potassium pump.
   a. true
   b. false

_____ 7. Rapid conduction of a nerve impulse in vertebrates is due to
   a. the large diameters of the axons.
   b. openings in the myelin sheath.
   c. an abundance of synapses.
   d. the high permeability of neuronal membranes to ions.
   e. All of these are correct.

_____ 8. What is involved in a nerve impulse?
   a. ions
   b. electrons
   c. atoms
   d. molecules

_____ 9. Synaptic vesicles are
   a. at the ends of dendrites and axons.
   b. at the ends of axons only.
   c. along the lengths of long fibers.
   d. All of these are correct.

_____ 10. Acetylcholine
   a. is a neurotransmitter.
   b. crosses the synaptic cleft.
   c. is broken down by acetylcholinesterase.
   d. All of these are correct.

_____ 11. A spinal nerve is a
   a. motor nerve.
   b. sensory nerve.
   c. mixed nerve.
   d. All of these are correct.

_____ 12. Automatic responses to specific external stimuli require
   a. rapid impulse transmission along the spinal cord.
   b. the involvement of the brain.
   c. simplified pathways called reflex arcs.
   d. the involvement of the autonomic system.

_____ 13. Which portion of the nervous system is required for a reflex arc?
   a. mixed spinal nerve
   b. gray matter of spinal cord
   c. cerebrum
   d. Both _a_ and _b_ are correct.
   e. _a_, _b_, and _c_ are correct.

_____ 14. The autonomic system has two divisions called the
   a. CNS and peripheral nervous system.
   b. somatic and skeletal systems.
   c. efferent and afferent.
   d. sympathetic and parasympathetic.

_____ 15. Motor axons of the somatic system release
   a. acetylcholine.
   b. norepinephrine.
   c. dopamine.
   d. serotonin.

_____ 16. Which system is active during a stressful time?
   a. parasympathetic
   b. sympathetic
   c. somatic
   d. All of these are correct.

_____ 17. The neurotransmitter of the parasympathetic system is
   a. norepinephrine.
   b. acetylcholine.
   c. cholinesterase.
   d. Both _a_ and _b_ are correct.

_____ 18. Which is the largest part of the human brain?
   a. cerebrum
   b. cerebellum
   c. medulla
   d. thalamus

_____ 19. The function of the cerebellum is
   a. consciousness.
   b. motor coordination.
   c. homeostasis.
   d. sensory reception.

_____ 20. Which portion of the brain is involved in judgment?
   a. cerebellum
   b. frontal lobe of cerebrum
   c. medulla oblongata
   d. parietal lobe of cerebrum

_____ 21. The drug that is classified as a hallucinogen is
   a. marijuana.
   b. alcohol.
   c. caffeine.
   d. nicotine.

_____ 22. Drugs of abuse primarily affect the
   a. cerebellum.
   b. medulla oblongata.
   c. limbic system.
   d. thalamus.

_____ 23. A ready-to-smoke, highly addictive form of cocaine is
   a. heroin.
   b. crack.
   c. marijuana.
   d. alcohol.

Answer in complete sentences.

24. Contrast the way the nerve impulse travels along an axon with the way it travels across a synapse.

25. In either of what two ways would you expect an inhibitory psychoactive drug to affect transmission across a synapse?

Test Results: _____ number correct ÷ 25 =_____ × 100 =_____%

ANSWER KEY

STUDY QUESTIONS

1. a. sends signal to cell body **b.** control center **c.** takes impulse away from cell body **2. a.** sensory receptor **b.** cell body **c.** axon **d.** dendrites **e.** axon **f.** dendrites **g.** axon **h.** nucleus of Schwann cell **i.** node of Ranvier **j.** effector **3.** to take nerve impulses to CNS **4.** to take nerve impulses from one part of CNS to another **5.** to take nerve impulses away from CNS **6. a.** an instrument that measures potential differences in millivolts **b.** Large organic ions are inside the membrane (also there is a slow leakage of K^+ to outside). **c.** more Na^+ outside than inside **d.** less K^+ outside than inside **e.** sodium-potassium pump **7. a.** depolarization **b.** repolarization **c.** action potential **d.** threshold **e.** resting potential **8.** sodium-potassium **9. a.** axon terminals **b.** synaptic vesicles **c.** dendrite **d.** neurotransmitter **e.** synaptic cleft **10.** reception of neurotransmitter at receptor site **11. a.** F **b.** T **c.** F **d.** T **e.** F **12. a.** dorsal root **b.** white matter **c.** gray matter **d.** ventral root **e.** central canal **13. a.** F, . . . contains myelinated axons . . . **b.** T **c.** F, . . . and a loss of voluntary control . . . **14. a.** cerebellum **b.** medulla oblongata **c.** pons **d.** pituitary gland **e.** cerebrum **f.** corpus callosum **g.** thalamus **15. a.** motor control, higher levels of thought **b.** integrates and sends sensory information to cerebrum **c.** homeostasis **d.** motor coordination **e.** control of internal organs **16. a.** frontal lobe **b.** parietal lobe **c.** occipital lobe **d.** temporal lobe **e.** all lobes **f.** frontal lobe **17.** a, b, c **18.** d **19. a.** cortical sensory areas **b.** prefrontal area **c.** diencephalon **d.** cortical sensory areas **e.** prefrontal area **f.** The hippocampus serves as bridge between prefrontal area and cortical sensory areas. **g.** The amygdala adds emotional overtones to memories. **h.** One set of arrows is for semantic memory and one set for episodic memory. **20. a.** T **b.** T **c.** T **21. a.** cranial **b.** spinal **c.** bundle of fibers (axons) **d.** It contains both sensory and motor fibers. **22. a.** sensory receptor **b.** sensory neuron **c.** dorsal root ganglion **d.** interneuron **e.** motor neuron **f.** effector **23. a.** sensory receptors **b.** sensory neurons **c.** interneurons **d.** motor neurons **24.** Tracts in CNS take impulses up from the cord to the brain. **25. a.** control internal organs **b.** have motor neurons **c.** have ganglia

26. a. fight or flight **b.** normal activity **c.** norepinephrine (NE) **d.** acetylcholine (ACh) **e.** near cord **f.** near organ **g.** spinal nerves only **h.** spinal nerves plus vagus **27. a.** limbic system **b.** increased likelihood of excitation **c.** decreased likelihood of excitation **28. a.** nicotine **b.** marijuana **c.** alcohol **d.** Crack **29.** Body's production of endorphins has decreased.

SIMON SAYS ABOUT NERVOUS CONDUCTION

1. c **2.** d **3.** d **4.** d **5.** g **6.** b **7.** c **8.** d **9.** a **10.** d

DEFINITIONS WORDSEARCH

```
G A N G L I O N
    T H A L A M U S   C
N   L I M B I C Y      I
O               N      T
C E R E B E L L U M A  A
  U X E L F E R   P    M
D E N D R I T E   S    O
N               N      E S
A C E T Y L C H O L I N E
```

a. thalamus **b.** neuron **c.** cerebellum **d.** dendrite **e.** reflex **f.** synapse **g.** limbic **h.** acetylcholine **i.** somatic **j.** ganglion

CHAPTER TEST

1. d **2.** b **3.** c **4.** c **5.** d **6.** a **7.** b **8.** a **9.** b **10.** d **11.** c **12.** c **13.** d **14.** d **15.** a **16.** b **17.** b **18.** a **19.** b **20.** b **21.** a **22.** c **23.** b **24.** There is an exchange of Na^+ and K^+ as the nerve impulse travels along an axon, but the release of a neurotransmitter causes the nerve impulse to travel across a synapse. **25.** An inhibitory psychoactive drug could either prevent the action of an excitatory neurotransmitter or promote the action of an inhibitory neurotransmitter at a synapse.

13

SENSES

Sensory receptors are specialized structures designed to convey information about the environment to the central nervous system. They help our bodies maintain homeostasis.

In this chapter, you will first learn about the interesting variety of receptors in the human body (p. 244). The skin has receptors that respond to four different types of stimuli (pp. 246–247). The special senses, associated with the head, are the most complex and fascinating of the senses. The senses of taste and smell both employ chemoreceptors (pp. 248–249) to detect the quality of food and to determine other characteristics of the environment. Eyes are intriguing organs, designed to perceive light (pp. 250–253). Vision is possible because of visual pigments (p. 252) of photoreceptors that become altered when light strikes them, generating a nervous impulse.

When you study the eye, make a diagram like Figure 13.6, and label all the parts of the eye. Next, make note of the functions of each part of the eye (p. 250). Pay close attention to how the retina is structured (pp. 251, 253) because that will help you understand why photoreceptors work the way they do (p. 252).

The ear houses the senses of hearing and equilibrium, making use of mechanoreceptors in both cases. Sketch the anatomy of the ear (see Fig. 13.12), labeling the parts and their functions. Be sure to include the anatomy and functioning of the inner ear (see Fig. 13.13), where hair cells for hearing and equilibrium are located.

TIP: Be sure to visit the Online Learning Center that accompanies *Human Biology* 9/e. It has practice quizzes, interactive activities, labeling exercises, art quizzes, animations, flash cards, and much more.
http://www.mhhe.com/maderhuman9

STUDY QUESTIONS

Study the text section by section. Answer the study questions so that you can fulfill the learning objectives for each section.

13.1 SENSORY RECEPTORS AND SENSATIONS (PAGES 244–245)

After you have answered the questions for this section, you should be able to
- Describe the various receptors, and state the type of stimuli they receive.
- Explain how sensation occurs.

1. Match the descriptions to these terms:

   chemoreceptors   mechanoreceptors   proprioceptors   thermoreceptors   pain receptors   photoreceptors

   a. _____ located only in the eye

   b. _____ monitor the pH of the blood

   c. _____ detect tissue damage

   d. _____ detect stretch in tendons and ligaments

   e. _____ sensitive to changes in heat and cold

   f. _____ hearing

   g. _____ taste and smell

Indicate whether the following statements are true (T) or false (F). Rewrite the false statements to make true statements.

2. _____ When stimulated, a sensory receptor generates a nerve impulse that travels in a sensory neuron to the CNS. Rewrite: _____

3. _____ Usually, each type of sensory receptor is sensitive to only one stimulus. Rewrite: _____

_____

4. _____ Sensation occurs in the brain and not at the sensory receptor. Rewrite: _____

_____

5. _____ Sensory receptors generate nerve impulses. Rewrite: _____

_____

6. _____ Sensory receptors are part of a reflex arc. Rewrite: _____

_____

13.2 PROPRIOCEPTORS AND CUTANEOUS RECEPTORS (PAGES 246–247)

After you have answered the questions for this section, you should be able to
- Name two types of proprioceptors and describe their functions.
- Associate the names of cutaneous receptors with their functions.

7. Name two types of proprioceptors: a. _____ and b. _____.

When a muscle spindle relaxes, c. _____.

Muscle spindles have two functions, d. _____ and e. _____.

8. Match the types of cutaneous receptors with the sense they are associated with.
   a. pressure   b. heat   c. cold   d. pain   e. touch

   _____ free nerve endings
   _____ Merkel disks
   _____ Krause end bulbs
   _____ Meissner corpuscles
   _____ Pacinian corpuscles
   _____ Ruffini endings

13.3 SENSES OF TASTE AND SMELL (PAGES 248–249)

After you have answered the questions for this section, you should be able to
- Describe the senses that rely on chemoreceptors.
- Explain how chemoreceptors operate.

9. Match the descriptions to these terms:

   taste receptors   smell receptors   both taste and smell receptors

   a. _____ receptor proteins combine with chemicals
   b. _____ brain senses impulses as a weighted average
   c. _____ microvilli house receptor proteins
   d. _____ salty receptor proteins on tip of the tongue
   e. _____ olfactory cells
   f. _____ are not effective when you have a cold
   g. _____ easily adapt to outside stimuli
   h. _____ involved in enjoyment of food
   i. _____ a characteristic combination of receptor proteins are activated

10. The senses of taste and smell work because specific ^{a.}_____ in the organs of taste and smell

combine with ^{b.}_____ in the air or food. Both senses thus employ ^{c.}_____ to

detect changes in the environment.

13.4 SENSE OF VISION (PAGES 250–255)

After you have answered the questions for this section, you should be able to
- Describe the anatomy of the eye and the function of each part.
- Describe the receptors for sight, their mechanism of action, and the mechanism for stereoscopic vision.
- Identify common disorders of sight discussed in the text.

11. In the following diagram, label the parts of the eye using the following alphabetized list of terms. Using the answer blanks provided, state the name and function of each part of the eye indicated in the illustration.

choroid
ciliary body
cornea
fovea centralis
iris
lens
optic nerve
retina
sclera

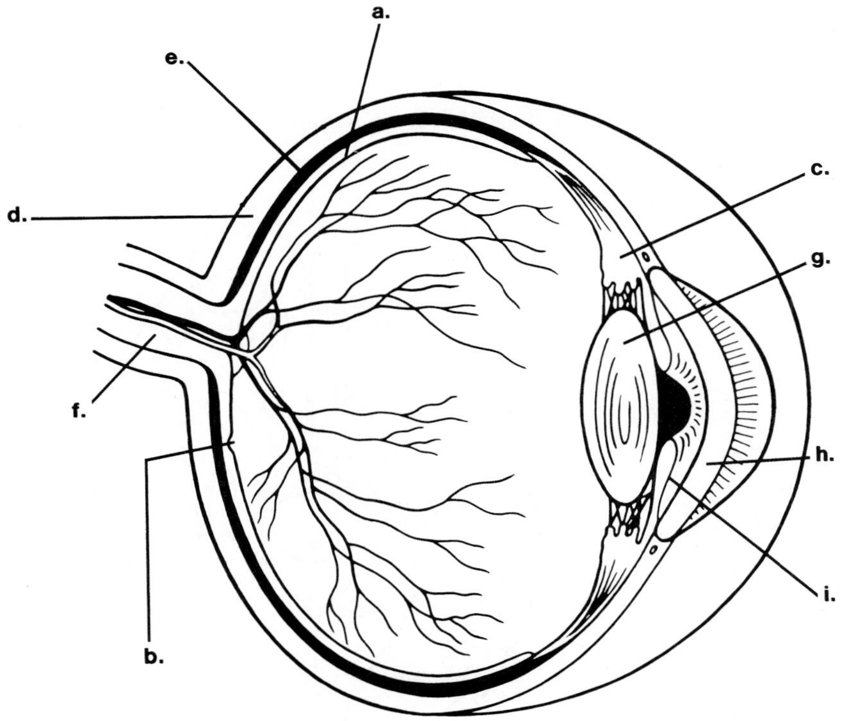

Structure   **Function**

a. _____   _____

b. _____   _____

c. _____   _____

d. _____   _____

e. _____   _____

f. _____   _____

g. _____   _____

h. _____   _____

i. _____   _____

In questions 12–16, fill in the blanks.

12. The lens is ^{a.}_____ for distant objects and ^{b.}_____ for close objects.

This is called ^{c.}_____.

13. The receptors of sight are classified as ^{a.}_____. Two types exist: ^{b.}_____ perceive

motion and are responsible for night vision, and ^{c.}_____ are responsible for color vision.

14. Rod cells have a pigment called a._____, made up of the protein b._____ and a pigment molecule called c._____, a derivative of vitamin A. When light strikes the pigment molecule, the rhodopsin is activated. Color vision depends on d._____ types of cones, each of which has a slightly different structure of e._____ molecule. Each is able to detect a different wavelength, or color, of light. Rod cells are located throughout the retina, but cone cells are concentrated in the f._____.

15. Considering the layers of the retina—rod cells and cone cells/bipolar cells/ganglion cells—which of these are at the back of the retina (closest to the choroid)? a._____ Which of these are receptors for sight? b._____ Which of these are fewest in number? c._____ This supports the belief that d._____ occurs in the retina.

16. With reference to the following figure, the region where the optic nerves cross is the a._____. Each primary visual area of the cerebral cortex receives information about b._____ (*the complete* or *one-half the*) visual field. Also, it is now known that the visual areas c._____ (*parcel out* or *retain as a unified whole*) information regarding color, form, motion, and so on. This means that the cerebral cortex has to d._____ (*rebuild* or *imagine*) the visual field before we can "see" it.

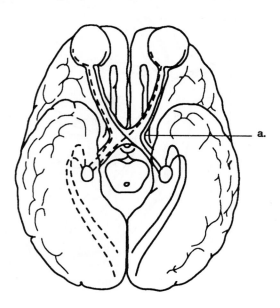

17. Fill in the blanks in this table.

Name	Description	Image Focused	Correction
Nearsightedness	See nearby objects	a.	Concave lens
Farsightedness	b.	c.	d.
Astigmatism	Cannot focus	Image not focused	e.

13.5 SENSE OF HEARING (PAGES 256–257)

After you have answered the questions for this section, you should be able to
- Describe the anatomy of the ear and the function of each part.
- Describe the sequence of events leading to an auditory nerve impulse.

18. Using the following alphabetized list of terms, label the parts of the ear. Using the answer blanks provided, state the name and function of each part of the ear indicated in the illustration.

auditory canal
auditory tube
cochlea
cochlear nerve
malleus (hammer)
pinna
semicircular canal
stapes (stirrup)
tympanic membrane
vestibule

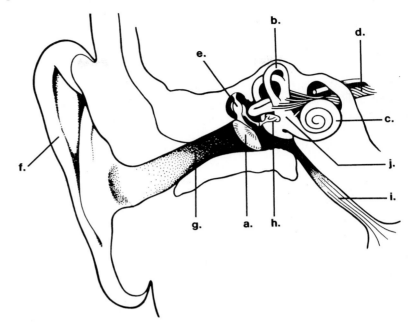

Structure/Function

a. _____
c. _____
e. _____
g. _____
i. _____

Structure/Function

b. _____
d. _____
f. _____
h. _____
j. _____

19. What is the proper sequence of the events that lead to the formation of an auditory nerve impulse? Indicate by letter. _____
   a. Vibration is transferred from the malleus to the incus to the stapes.
   b. Basilar membrane moves up and down.
   c. Nerve impulse is transmitted in the cochlear nerve to brain.
   d. Sound waves pass through the auditory canal.
   e. Cilia of hair cells rub against the tectorial membrane.
   f. Sound waves cause the tympanic membrane to vibrate.
   g. Nerve impulse is generated.
   h. Vibrations move from the vestibular canal to the tympanic canal.
   i. Membrane at the oval window vibrates.

13.6 SENSE OF EQUILIBRIUM (PAGE 261)

After you have answered the questions for this section, you should be able to
• Explain rotational equilibrium and gravitational equilibrium.

In question 20, fill in the blanks.

20. a. _____ equilibrium is called for when the entire body is moving. Fluid in the semicircular canals displaces the gelatinous material within the b. _____. For c. _____ equilibrium, movement occurs in one plane (vertical or horizontal), and the otoliths in the d. _____ and saccule are displaced in the gelatinous material, bending the cilia of the hair cells. Hair cells in the ampullae and utricle and saccule synapse with the e. _____ nerve.

Review key terms by using the following alphabetized list of terms to fill in the blanks. Then complete the wordsearch.

```
A C C O N E C E L L I O N X
M R E H T O P S D N I L B V
P E T U O R E B Y U J D F E
U J I K S R R E D S I C L I
L H U T S C O C H L E A L O
L G Y C I S X I W A S N I U
A B H I C L O P D S I D E D
K O T O L I T H B C Z T Y U
M U J K E R H O D O P S I N
```

ampulla
blind spot
choroid
cochlea
cone cell
ossicle
otolith
rhodopsin

a. _____ Photoreceptor that responds to bright light.

b. _____ Vascular, pigmented layer of eyeball.

c. _____ Portion of inner ear that resembles a snail-shell in appearance.

d. _____ Area of eye where optic nerve passes through retina.

e. _____ Base of semicircular canal in inner ear.

f. _____ One of the small bones of the middle ear.

g. _____ Calcium carbonate granule within inner ear.

h. _____ Visual pigment found in rod cells.

CHAPTER TEST

OBJECTIVE TEST

Do not refer to the text when taking this test.

____ 1. A receptor
    a. is the first portion of a reflex arc.
    b. initiates a nerve impulse.
    c. usually responds to only one type of stimulus.
    d. is attached to a dendrite.
    e. All of the above are correct.

____ 2. There are five sensory receptors in the skin. They are
    a. choroid, cochlea, ossicles, and otoliths.
    b. hot, cold, pressure, touch, and pain.
    c. sclera, choroid, cornea, cones, and cochlea.
    d. mechanoreceptors, chemoreceptors, and photoreceptors.

____ 3. Taste cells and olfactory cells are both
    a. somatic senses.
    b. mechanoreceptors.
    c. pseudociliated epithelium.
    d. chemoreceptors.

____ 4. The blind spot is
    a. a nontransparent area on the lens.
    b. a nontransparent area on the cornea.
    c. on the retina, where there are no rods or cones.
    d. called the fovea centralis.

In questions 5–8, match the questions to these terms:
    a. retina b. optic nerve
    c. lens d. all e. none

____ 5. Which of these is (are) necessary to proper vision?

____ 6. Which of these contain(s) receptors for sight?

____ 7. Which of these focus(es) light?

____ 8. In which is the sensation of sight realized?

____ 9. The current theory of color vision proposes that
    a. there are three primary colors associated with color vision.
    b. cone cells respond selectively to different wavelengths of light.
    c. the rod cells are responsible for nighttime color vision.
    d. Both *a* and *b* are correct.
    e. *a, b,* and *c* are correct.

____ 10. Each side of the brain receives information from both eyes due to the
    a. optic chiasma.
    b. fovea centralis.
    c. ciliary muscle.
    d. ciliary body.

_____ 11. If you are nearsighted, the image is focused
  a. in front of the retina.
  b. behind the retina.
  c. on the retina.
  d. at the blind spot.
_____ 12. The disorders of nearsightedness and farsightedness are due to
  a. an eyeball of incorrect length.
  b. a cloudy lens.
  c. pressure increase.
  d. a torn retina.
_____ 13. Vitamin A is needed for the
  a. lens.
  b. rod cells.
  c. cone cells.
  d. cornea.
_____ 14. The cochlear nerve is associated with the
  a. spiral organ.
  b. ossicles.
  c. tympanic membrane.
  d. auditory tube.

In questions 15–17, match the questions to the terms.
  a. ossicles   b. otoliths
  c. cochlea   d. auditory tube

_____ 15. Which of these has (have) nothing to do with hearing?
_____ 16. In which would you find receptors for hearing?
_____ 17. Which of these is (are) concerned with balance?

_____ 18. Equilibrium receptors differ from hearing receptors in that equilibrium receptors _____, while hearing receptors _____.
  a. are located in the outer ear; are located in the inner ear
  b. respond to pressure waves; do not respond to pressure waves
  c. consist of hair cells; do not consist of hair cells
  d. are in the semicircular canals; are in the spiral organ
_____ 19. Which part of the ear is for equilibrium?
  a. outer
  b. middle
  c. inner
  d. All of these are correct.
_____ 20. Nerve deafness may be due to
  a. fused ossicles.
  b. worn stereocilia.
  c. German measles.
  d. Both *a* and *c* are correct.

In questions 21–25, match the questions to these terms.
  a. spiral organ   b. rods and cones
  c. pressure receptor   d. olfactory receptor
  e. taste buds   f. utricle and saccule

_____ 21. Which of these is (are) NOT mechanoreceptors?
_____ 22. Which of these is (are) located in the skin?
_____ 23. Which of these helps maintain equilibrium?
_____ 24. Which of these is (are) located in the ear?
_____ 25. Which of these is (are) stimulated by chemicals?

THOUGHT QUESTIONS

Answer in complete sentences.
26. The length of the spiral organ (organ of Corti) is related to what ability in humans?

27. Relate the function of the rod cells to their placement in the retina.

Test Results: _____ number correct ÷ 27 = _____ × 100 = _____%

ANSWER KEY

STUDY QUESTIONS

1. a. photoreceptors **b.** chemoreceptors **c.** pain receptors **d.** proprioceptors, mechanoreceptors **e.** thermoreceptors **f.** mechanoreceptors **g.** chemoreceptors **2.** T **3.** T **4.** T **5.** T **6.** T **7. a.** muscle spindle **b.** Golgi tendon organ **c.** nerve impulses are generated **d.** maintain posture **e.** tell position of limbs **8.** b, c and d, e, a, e, a,a **9. a.** both **b.** taste receptors **c.** taste receptors **d.** taste receptors **e.** smell receptors **f.** smell receptors **g.** smell receptors **h.** both **i.** smell receptors **10. a.** protein receptors **b.** chemicals (molecules) **c.** chemoreceptors **11. a.** retina; photoreceptors for sight **b.** fovea centralis; makes acute vision possible **c.** ciliary body; holds lens in place; accommodation **d.** sclera; protects eyeball **e.** choroid; absorbs stray light rays **f.** optic nerve; transmission of nerve impulse **g.** lens; focusing **h.** cornea; refracts light rays **i.** iris; regulates entrance of light **12. a.** flat **b.** rounded **c.** accommodation **13. a.** photoreceptors **b.** rod cells **c.** cone cells **14. a.** rhodopsin **b.** opsin **c.** retinal **d.** three **e.** opsin **f.** fovea centralis **15. a.** rod cells and cone cells **b.** rod cells and cone cells **c.** ganglion cells **d.** integration **16. a.** optic chiasma **b.** one-half the **c.** parcel out **d.** rebuild **17. a.** in front of retina **b.** see distant objects **c.** behind retina **d.** convex lens **e.** irregular lens **18. a.** tympanic membrane; starts vibration of ossicles **b.** semicircular canal; rotational equilibrium **c.** cochlea; contains mechanoreceptors for hearing **d.** cochlear nerve; transmission of nerve impulse **e.** malleus (hammer); transmits vibrations **f.** pinna; reception of sound waves **g.** auditory canal; collection of sound waves **h.** stapes (stirrup); transmits vibrations to oval window **i.** auditory tube; connects middle ear to pharynx **j.** vestibule; gravitational equilibrium **19.** d, f, a, i, h, b, e, g, c **20. a.** rotational **b.** cupula **c.** gravitational **d.** utricle **e.** vestibular

DEFINITIONS WORDSEARCH

```
A    C O N E C E L L
M       H T O P S D N I L B
P       O
U       S R
L       S C O C H L E A
L       I     I
A       C     D
  O T O L I T H
    E R H O D O P S I N
```

a. cone cell **b.** choroid **c.** cochlea **d.** blind spot **e.** ampulla **f.** ossicle **g.** otolith **h.** rhodopsin

CHAPTER TEST

1. e **2.** b **3.** d **4.** c **5.** d **6.** a **7.** c **8.** e **9.** d **10.** a **11.** a **12.** a **13.** b **14.** a **15.** b and d **16.** c **17.** b **18.** d **19.** c **20.** b **21.** b, d, e **22.** c **23.** f **24.** a, f **25.** d, e **26.** The length of the spiral organ (organ of Corti) is related to the ability of humans to hear a range of pitches. **27.** The rod cells are located throughout the retina, and this is consistent with their ability to detect motion.

14

ENDOCRINE SYSTEM

STUDY QUESTIONS

Study the text section by section. Answer the study questions so that you can fulfill the learning objectives for each section.

14.1 ENDOCRINE GLANDS (PAGES 266–269)

After you have answered the questions for this section, you should be able to
- State the location of endocrine glands, and label an illustration that shows the endocrine glands.
- Associate the name of the gland with its hormones.
- Associate hormones with their actions.
- Tell in general how chemical signals influence the behavior and/or metabolism of cells.
- Distinguish between the effects of steroid and peptide hormones on cells.

1. Label the endocrine glands in the following diagram, and name at least one hormone produced by each.

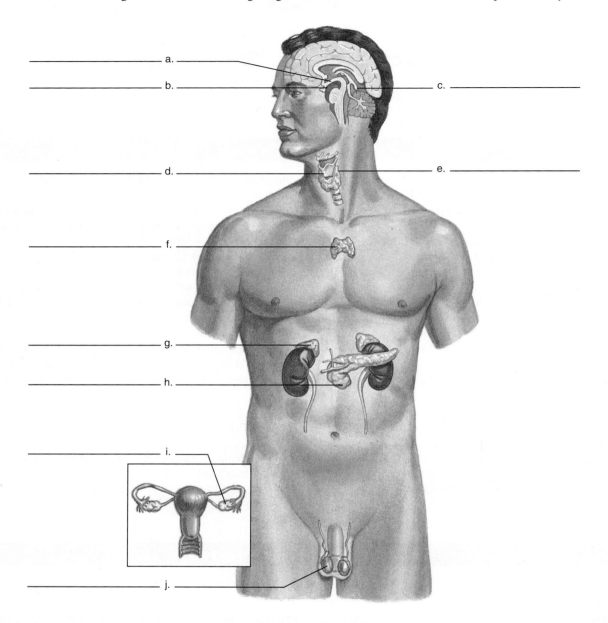

_____ a. _____

_____ b. _____   c. _____

_____ d. _____   e. _____

_____ f. _____

_____ g. _____

_____ h. _____

_____ i. _____

_____ j. _____

2. The hormone ᵃ·_____ is released by the pineal gland. Its target organ is the
ᵇ·_____, where it controls ᶜ·_____ rhythms.

3. Write either *peptide hormone* or *steroid hormone* on the lines above each diagram. Using the following alphabetized list of terms, place an appropriate word or phrase on the lines within each diagram.

cAMP   hormone-receptor complex   protein synthesis   receptor protein

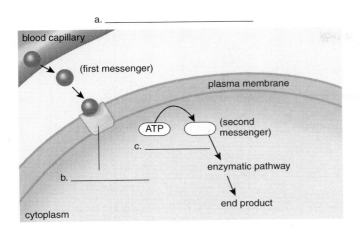

a. _____

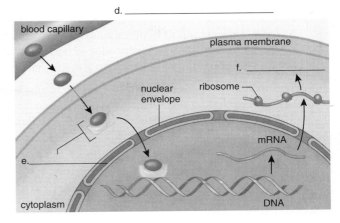

d. _____

14.2 HYPOTHALAMUS AND PITUITARY GLAND (PAGES 270–272)

After you have answered the questions for this section, you should be able to
- Contrast the ways in which the hypothalamus controls the posterior and anterior pituitary.
- List the hormones produced by the posterior and anterior pituitary, and describe their effects.

4. Match the descriptions to the numbers in the following diagram.
   a. _____ hypothalamus
   b. _____ anterior pituitary
   c. _____ target gland
   d. _____ feedback that inhibits hypothalamus
   e. _____ feedback that inhibits anterior pituitary
   f. _____ releasing hormone
   g. _____ stimulating hormone
   h. _____ target gland hormone

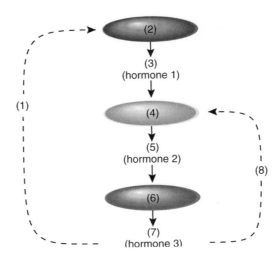

5. The hypothalamus controls a._____, b._____, and c._____. It

   also controls the secretions of the d._____.

6. Place the appropriate letters next to each statement.
   AP—anterior pituitary   PP—posterior pituitary
   a. _____ connected to hypothalamus by nerve fibers
   b. _____ connected to hypothalamus by blood vessels
   c. _____ secretes hormones produced by hypothalamus
   d. _____ controlled by releasing hormones produced by hypothalamus

7. To show why the anterior pituitary is sometimes called the master gland, complete this table.

Anterior Pituitary Produces	Gland Controlled	Hormone Produced by Gland
TSH	a.	b.
ACTH	c.	d.
Gonadotropic hormones		
Female	e.	f.
Male	g.	h.

8. The anterior pituitary produces three other hormones. The hormone a._____ causes the mammary

   glands to develop and produce milk. b._____ hormone causes skin-color changes in many fishes,

   amphibians, and reptiles. Growth hormone (GH) promotes cell division, protein synthesis, and

   c._____ growth. If too little GH is produced during childhood, the individual exhibits pituitary

   d._____. If too much is produced, the individual can exhibit e._____. If there is an

   overproduction of GH in the adult, f._____ results, and the face, hands, and feet g._____.

14.3 THYROID AND PARATHYROID GLANDS (PAGES 273–274)

After you have answered the questions for this section, you should be able to
• List the hormones of the thyroid gland, and give their functions.
• Give the function of parathyroid hormone.
• Identify thyroid disorders.

9. Match the phrases to these conditions:
   cretinism   exophthalmic goiter   simple goiter   myxedema
   a. _____ hypothyroidism (choose more than one)
   b. _____ hyperthyroidism
   c. _____ hypothyroidism since birth
   d. _____ hypothyroidism in the adult
   e. _____ lack of iodine

10. Match the descriptions to these phrases:
    1. low blood Ca^{2+} 2. high blood Ca^{2+}
    3. Ca^{2+} is deposited in bones   4. Ca^{2+} is deposited in blood
    a. _____ Calcitonin is present.
    b. _____ Parathyroids are mistakenly removed during an operation.
    c. _____ Calcitonin will be released.
    d. _____ PTH will be released.
    e. _____ PTH is present.
    f. _____ Calcitonin will not be released.
    g. _____ PTH will not be released.

14.4 ADRENAL GLANDS (PAGES 275–277)

After you have answered the questions for this section, you should be able to
- Describe the hormones of the adrenal medulla, their control, and their effects.
- List the hormones of the adrenal cortex, state their functions, and tell how their release is controlled.
- Describe what happens when the adrenal cortex malfunctions.

11. Place the appropriate letters next to each description.
    AM—adrenal medulla   AC—adrenal cortex
    a. _____ inner portion of adrenal gland
    b. _____ outer portion of adrenal gland
    c. _____ hypothalamus sends nervous impulses
    d. _____ hypothalamus sends releasing hormone to anterior pituitary, and anterior pituitary sends ACTH to target gland
    e. _____ releases glucocorticoids and mineralocorticoids
    f. _____ releases epinephrine and norepinephrine
    g. _____ short-term response to stress
    h. _____ long-term response to stress

12. Distinguish between cortisol and aldosterone by writing *yes* or *no* on each line.

	Cortisol	Aldosterone
Controlled by ACTH	a._____	_____
Glucocorticoid	b._____	_____
Mineralocorticoid	c._____	_____
Na^+/K^+ balance	d._____	_____
Amino acids $\rightarrow$ glucose	e._____	_____
Controlled by angiotensin II	f._____	_____

13. When there is low blood Na^+, the kidneys secrete a._____ , an enzyme that converts angiotensinogen to b._____ , which later becomes c._____ in lung capillaries. The latter causes the adrenal cortex to release d._____ . Blood pressure now e._____ . The heart releases a hormone called f._____ that is antagonistic to aldosterone.

14. Place the appropriate letters next to these symptoms:
    AD—Addison disease   CS—Cushing syndrome
    a. _____ glucose cannot be replenished in times of stress
    b. _____ loss of sodium and water
    c. _____ tendency toward diabetes mellitus
    d. _____ low blood pressure because of a low blood sodium level
    e. _____ high blood pressure; masculization; weight gain
    f. _____ children show obesity and poor growth in height
    g. _____ bronzing of skin
    h. _____ obese trunk

After you have answered the questions for this section, you should be able to
- Describe the functions of insulin and glucagon.
- Discuss the problems associated with diabetes.

15. Write the word *insulin* or *glucagon* on the appropriate arrow.

    glycogen ^{a.} ←——————— glucose molecules

    glucose storage in liver ^{b.} ——————→ glucose in the blood

16. Complete each of the following statements with the term *increases* or *decreases*.

    Glucagon ^{a.}_____ blood sugar concentration. In type I diabetes, insulin production from the

    pancreas ^{b.}_____. In type II diabetes, the response of body cells to the influence of

    insulin ^{c.}_____.

After you have answered the questions for this section, you should be able to
- Describe the functions of the testes and ovaries, the thymus, and the pineal gland.
- Describe the function of leptin, growth factors, and prostaglandins, produced by various tissues as opposed to glands.

17. Match these numbered items to the glands. There is more than one match for each gland, and answers may be used more than once.

    1. T lymphocytes   2. melatonin   3. testosterone   4. circadian rhythm   5. males
    6. females   7. estrogen and progesterone   8. thymosins   9. secondary sex characteristics

    a. _____ testes
    b. _____ ovaries
    c. _____ thymus
    d. _____ pineal gland

18. Match these phrases to the chemical signals listed. Answers can be used more than once.

    1. cause(s) cell division   2. cause(s) a feeling of satiety   3. has (have) various effects   4. act(s) locally

    a. _____ leptin
    b. _____ growth factors
    c. _____ prostaglandins

After you have answered the questions for this section, you should be able to
- Give examples to show that both the endocrine system and nervous system utilize chemical signals.
- Tell how the endocrine system works with other systems of the body to maintain homeostasis.

19. Place the appropriate letters next to each description.
    ES—endocrine system   NS—nervous system
    a. _____ uses chemical signals that bind to protein receptors
    b. _____ sometimes acts from a distance and sometimes acts locally
    c. _____ utilizes releasing hormones
    d. _____ utilizes neurotransmitters
    e. _____ utilizes hormones that are distributed in the blood

20. The *Human Systems Work Together* diagram in your textbook shows how the endocrine system benefits from the actions of the other organ systems of the body. Match the organ systems to the descriptions.
    (1) Lymphatic vessels pick up excess tissue fluid; immune system protects against infections.
    (2) Bones provide protection for glands and store Ca^{2+}.
    (3) Kidneys keep blood values within normal limits so that transport of hormones continues.

(4) Stomach and small intestines produce hormones.
(5) Gas exchange in lungs provides oxygen and rids body of carbon dioxide.
(6) Hypothalamus is part of endocrine system; nerves innervate glands of secretion.
(7) Gonads produce sex hormones.
(8) Muscles help protect glands.
(9) Skin provides sensory input that results in the activation of endocrine glands.
(10) Blood vessels transport hormones from glands; blood services glands; heart produces atrial natriuretic hormones.

a. _____ integumentary system
b. _____ lymphatic system/immunity
c. _____ cardiovascular system
d. _____ skeletal system
e. _____ urinary system
f. _____ muscular system
g. _____ respiratory system
h. _____ reproductive system
i. _____ nervous system
j. _____ digestive system

 Hormone Hockey

For every five correct answers in sequence, you have scored one goal.

First goal: Match the hormone to the glands (a–i).
Glands:
a. anterior pituitary
b. thyroid
c. parathyroids
d. adrenal cortex
e. adrenal medulla
f. pancreas
g. gonads
h. pineal gland
i. posterior pituitary

Hormones:
1. ___ insulin
2. ___ oxytocin
3. ___ melatonin
4. ___ cortisol
5. ___ thyroxine

Second goal: Match the condition to the glands (a–i).
Conditions:
6. ___ diabetes mellitus
7. ___ cretinism
8. ___ Addison disease
9. ___ hypertension
10. ___ giantism

Third goal: Match the function to the hormones (a–l).
a. melatonin
b. estrogens
c. androgens
d. insulin
e. glucagon
f. epinephrine
g. aldosterone
h. cortisol
i. parathyroid hormone
j. thyroxine
k. calcitonin (lowers)
l. antidiuretic hormone

Functions:
11. ___ raises blood calcium level
12. ___ reduces stress
13. ___ maintains secondary female sex characteristics
14. ___ involved in circadian rhythms
15. ___ stimulates water reabsorption by kidneys

Fourth goal: Match the glands to the hormones (a–l). Some glands require two answers.

Glands:
16. ___ testes
17. ___ adrenal cortex
18. ___ pancreas
19. ___ thyroid
20. ___ adrenal medulla

Fifth goal: Select five hormones secreted by the anterior pituitary by answering yes or no to each of these.

21. ___ thyroid-stimulating hormone
22. ___ androgens
23. ___ gonadotropic hormones
24. ___ glucagon
25. ___ oxytocin
26. ___ growth hormone
27. ___ prolactin
28. ___ antidiuretic hormone
29. ___ estrogens
30. ___ adrenocorticotropic hormone

How many goals did you make? _____

DEFINITIONS WORDSEARCH

Using the following alphabetized list of terms, review key terms by completing this wordsearch.

```
C R E T I N I S M N U J L I P
O A R H G T Z B E S A I L H R
R A L P O X Y T O C I N H U O
T M L C F R S H U K N H S R S
I E R G I U M N V Y S A F T T
S D G T R T D O F W U J H U A
O E P O L I O U N Y L R E R G
L X H Y T F G N B E I S E F L
P Y J T D H S E I D N S G B A
I M P H E R O M O N E B G F N
N O R E P I N E P H R I N E D
S G W Y T J T U D H R T K F I
P T Y E R W Q A S E D F R G N
```

calcitonin
cortisol
cretinism
hormone
insulin
myxedema
norepinephrine
oxytocin
prostaglandin

a. _____ Thyroid hormone that regulates the blood calcium level.

b. _____ Chemical signal.

c. _____ Condition due to improper development of thyroid in infants.

d. _____ Posterior pituitary hormone causing uterine contractions and milk letdown.

e. _____ Pancreatic hormone that lowers the blood glucose level.

f. _____ Adrenal gland hormone that increases the blood glucose level.

g. _____ Condition caused by lack of thyroid hormone in adult.

h. _____ Stress hormone from adrenal medulla.

i. _____ Local tissue hormone.

CHAPTER TEST

OBJECTIVE TEST

Do not refer to the text when taking this test.

____ 1. All hormones are believed to
   a. have plasma membrane receptors.
   b. affect cellular metabolism.
   c. increase the amount of cAMP.
   d. increase the amount of protein synthesis.

____ 2. The adrenal glands are
   a. at the base of the brain.
   b. on the trachea.
   c. on the kidneys.
   d. beneath the stomach.

____ 3. Which statement is NOT true about hormones?
   a. Hormones search throughout the bloodstream for their receptors.
   b. They act as chemical signals.
   c. They are released by endocrine glands.
   d. They can affect our appearance, our metabolism, or our behavior.

____ 4. Hormonal secretions are most often controlled by
   a. negative feedback mechanisms.
   b. positive feedback mechanisms.
   c. the hormone insulin.
   d. the cerebrum of the brain.

____ 5. Steroid hormones
   a. combine with receptors in the plasma membrane.
   b. pass through the membrane.
   c. activate genes leading to protein synthesis.
   d. Both *b* and *c* are correct.

____ 6. Which gland produces the greatest number of hormones?
   a. posterior pituitary
   b. anterior pituitary
   c. thymus
   d. pineal gland

_____ 7. The hypothalamus controls the anterior pituitary via
    a. nervous stimulation.
    b. the midbrain.
    c. vasopressin.
    d. releasing hormones.

_____ 8. ADH and oxytocin are secreted by the
    a. hypothalamus.
    b. posterior pituitary.
    c. thyroid gland.
    d. parathyroids.

_____ 9. Which hormone is involved with milk production and nursing?
    a. prolactin
    b. androgens
    c. antidiuretic hormone
    d. growth hormone

_____ 10. The anterior pituitary stimulates the
    a. thyroid.
    b. adrenal cortex.
    c. adrenal medulla.
    d. pancreas.
    e. Both *a* and *b* are correct.

_____ 11. Too much urine matches too
    a. little ADH.
    b. much ADH.
    c. little ACTH.
    d. much ACTH.

_____ 12. Thyroxine
    a. increases metabolism.
    b. stimulates the thyroid gland.
    c. lowers oxygen uptake.
    d. All of these are correct.

_____ 13. The adrenal cortex produces hormones affecting
    a. glucose metabolism.
    b. amino acid metabolism.
    c. sodium balance.
    d. All of these are correct.

_____ 14. Which hormone regulates blood calcium levels?
    a. calcitonin
    b. parathyroid hormone
    c. cortisol
    d. Both *a* and *b* are correct.

_____ 15. Which gland produces sex hormones?
    a. anterior pituitary
    b. posterior pituitary
    c. adrenal cortex
    d. Both *a* and *b* are correct.

_____ 16. Tetany occurs when there is too
    a. little calcium in the blood.
    b. much calcium in the blood.
    c. little sodium in the blood.
    d. much sodium in the blood.

_____ 17. Cushing syndrome is due to a malfunctioning
    a. thyroid.
    b. adrenal cortex.
    c. adrenal medulla.
    d. pancreas.

_____ 18. A simple goiter is caused by
    a. too much salt in the diet.
    b. too little iodine in the diet.
    c. too many sweets in the diet.
    d. a bland diet.

_____ 19. Acromegaly might be due to a tumor of the
    a. pancreas.
    b. anterior pituitary.
    c. thyroid.
    d. adrenal cortex.

_____ 20. If a person is suffering from hypoglycemia, he or she should
    a. be given some sugar.
    b. sit with the head down.
    c. be given insulin.
    d. not eat fatty foods.

_____ 21. In which case is insulin not produced?
    a. type 1 diabetes
    b. type 2 diabetes
    c. type 3 diabetes
    d. type 4 diabetes

_____ 22. Which of these is NOT a similarity between the nervous and endocrine systems?
    a. Both use chemical signals.
    b. Both utilize hormones.
    c. Both regulate other organ systems.
    d. Both act from a distance or act locally.

_____ 23. Which one of these is NOT an endocrine gland?
    a. thymus gland
    b. prostate gland
    c. pineal gland
    d. testis and ovary

_____ 24. Which one of these is incorrect?
    a. Some hormones are produced by tissues/cells, not glands.
    b. Chemical signals encompass hormones and other types of molecules that act as messengers between cells, organs, and individuals.
    c. Peptide, but not steroid, hormones require a second messenger.
    d. All of these are correct.

_____ 25. Which one of these is received by a receptor protein in the plasma membrane?
    a. peptide hormone
    b. first messenger
    c. epinephrine
    d. All of these are correct.

Answer in complete sentences.

26. Explain the occurrence of a goiter when an individual does not receive enough iodine in the diet.

27. Why does the release of renin by the kidneys cause the blood pressure to rise?

Test Results: number correct ÷ 27 =   × 100 =   %

ANSWER KEY

STUDY QUESTIONS

1. a. hypothalamus, hypothalamic-releasing and inhibiting hormone **b.** pituitary gland, growth hormone, ACTH **c.** pineal gland, melatonin **d.** thyroid gland, thyroxine, calcitonin **e.** parathyroid, parathyroid hormone **f.** thymus, thymosin **g.** adrenal gland, cortisol, aldosterone, epinephrine, norepinephrine **h.** pancreas, insulin, glucagon **i.** ovary, estrogen, progesterone **j.** testis, testosterone **2. a.** melatonin **b.** brain **c.** circadian **3. a.** peptide hormone **b.** receptor protein **c.** cAMP **d.** steroid hormone **e.** hormone-receptor complex **f.** protein synthesis **4. a.** (2) **b.** (4) **c.** (6) **d.** (1) **e.** (8) **f.** (3) **g.** (5) **h.** (7) **5. a.** heartbeat **b.** body temperature **c.** H_2O balance **d.** pituitary gland **6. a.** PP **b.** AP **c.** PP **d.** AP **7. a.** thyroid **b.** thyroid hormones **c.** adrenal cortex **d.** cortisol **e.** ovaries **f.** sex hormones **g.** testes **h.** sex hormones **8. a.** prolactin **b.** Melanocyte-stimulating **c.** skeletal and muscular **d.** dwarfism **e.** gigantism **f.** acromegaly **g.** enlarge **9. a.** cretinism, simple goiter, myxedema **b.** exophthalmic goiter **c.** cretinism **d.** myxedema **e.** simple goiter **10. a.** (2) and (3) **b.** (1) **c.** (2) **d.** (1) **e.** (1) and (4) **f.** (1) **g.** (2) **11. a.** AM **b.** AC **c.** AM **d.** AC **e.** AC **f.** AM **g.** AM **h.** AC **12. a.** yes, no **b.** yes, no **c.** no, yes **d.** no, yes **e.** yes, no **f.** no, yes **13. a.** renin, **b.** angiotensin I **c.** angiotensin II **d.** aldosterone **e.** rises **f.** atrial natriuretic hormone (ANH) **14. a.** AD **b.** AD **c.** CS **d.** AD **e.** CS **f.** CS **g.** AD **h.** CS **15. a.** insulin **b.** glucagon **16. a.** increases **b.** decreases **c.** decreases **17. a.** 3, 5, 9 **b.** 6, 7, 9 **c.** 1, 8 **d.** 2, 4 **18. a.** 2 **b.** 1 **c.** 3, 4 **19. a.** ES, NS **b.** ES, NS **c.** ES **d.** NS **e.** ES **20. a.** (9) **b.** (1) **c.** (10) **d.** (2) **e.** (3) **f.** (8) **g.** (5) **h.** (7) **i.** (6) **j.** (4)

HORMONE HOCKEY

First goal: **1.** f **2.** i **3.** h **4.** d **5.** b. Second goal: **6.** f **7.** b **8.** d **9.** d **10.** a. Third goal: **11.** i **12.** h **13.** b **14.** a **15.** l Fourth goal: **16.** c **17.** g and h

18. d and e **19.** j **20.** f. Fifth goal: **21.** yes **22.** no **23.** yes **24.** no **25.** no **26.** yes **27.** yes **28.** no **29.** no **30.** yes

DEFINITIONS WORDSEARCH

```
C R E T I N I S M          P
O   A   H                  R
R A L   O X Y T O C I N    O
T M   C   R       N        S
I E   I   M       S        T
S D     T   O     U        A
O E       O   N   L        G
L X         N E I          L
  Y           I N          A
  M           N            N
N O R E P I N E P H R I N E D
                           I
                           N
```

a. calcitonin **b.** hormone **c.** cretinism **d.** oxytocin **e.** insulin **f.** cortisol **g.** myxedema **h.** norepinephrine **i.** prostaglandin

CHAPTER TEST

1. b **2.** c **3.** a **4.** a **5.** d **6.** b **7.** d **8.** b **9.** a **10.** e **11.** a **12.** a **13.** d **14.** d **15.** c **16.** a **17.** b **18.** b **19.** b **20.** a **21.** a **22.** b **23.** b **24.** d **25.** d **26.** When an individual does not receive enough iodine in the diet, the thyroid is unable to produce thyroxine. The lack of thyroxine in the blood causes the anterior pituitary to produce more TSH, and this hormone promotes an increase in the size of the thyroid. **27.** Renin leads to the formation of angiotensin II, which stimulates the adrenal cortex to secrete aldosterone. Aldosterone causes sodium to be reabsorbed by the kidneys, and this leads to an increase in blood volume and blood pressure.

PART 5 REPRODUCTION IN HUMANS

15

REPRODUCTIVE SYSTEM

The reproductive system allows us to reproduce ourselves and does not participate in maintaining homeostasis to the same extent as other systems. The sex hormones help us maintain health and vigor but are not critical toward that end.

Study the diagrams of the male (p. 290) and female (p. 294) reproductive tracts. Know the functions of each structure and how they participate in the reproductive process. Reproduction is under hormonal control, beginning with the hypothalamus. It triggers the release of FSH and LH from the anterior pituitary in both males

and females. These two hormones then influence the production of gametes as well as the production of the male and female hormones responsible for secondary sexual characteristics. Make a chart of the ovarian cycle of female hormones that control the uterine cycle during the monthly menstrual cycle (pp. 298–299).

This chapter closes with a discussion of methods of birth control (pp. 301–302). Study this information carefully. The method of birth control you choose may also be able to protect you from acquiring many sexually transmitted diseases, the subject of Chapter 23.

TIP: Be sure to visit the Online Learning Center that accompanies *Human Biology* 9/e. It has practice quizzes, interactive activities, labeling exercises, art quizzes, animations, flash cards, and much more. http://www.mhhe.com/maderhuman9

STUDY QUESTIONS

Study the text section by section. Answer the study questions so that you can fulfill the learning objectives for each section.

15.1 MALE REPRODUCTIVE SYSTEM (PAGES 290–293)

After you have answered the questions for this section, you should be able to
- Describe the structure and function of the male reproductive system.
- State the path sperm take from the site of production until they exit the male.
- Name the glands that add secretions to semen.
- Discuss hormonal regulation of sperm production in the male.
- Name the actions of testosterone, including both primary and secondary sex characteristics.

1. Using the alphabetized list of terms and the blanks provided, identify and state a function for the parts of the human male reproductive and urinary systems shown in the following diagram.

bulbourethral gland   epididymis   penis   prostate gland   seminal vesicles
testis   urethra   urinary bladder   vas deferens

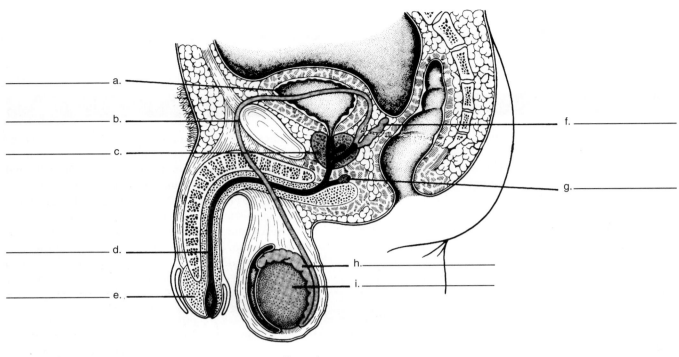

Structure	Function
a. _____	_____
b. _____	_____
c. _____	_____
d. _____	_____
e. _____	_____
f. _____	_____
g. _____	_____
h. _____	_____
i. _____	_____

2. Place the appropriate letters next to each phrase.

   ST—seminiferous tubules   IC—interstitial cells

   a. _____ produce androgens
   b. _____ produce sperm
   c. _____ controlled by FSH
   d. _____ controlled by LH

In questions 3–7, fill in the blanks.

3. Trace the path of sperm through the male reproductive system.

   Testes to the a._____ to the vas deferens to the b._____ to the c._____.

4. What three organs add secretions to seminal fluid?

   a._____

   b._____

   c._____

5. What is the general function of these secretions? _____

   _____

6. The process of sperm production is called a._____. This occurs inside

   b._____ tubules inside each testis.

7. Mature sperm cells have three parts: a._____, b._____, and

   c._____. A cap called the d._____ contains enzymes that allow a sperm to enter

   an egg. What section of a sperm contains the mitochondria that provide energy for motility?

   e._____

8. Indicate whether these statements are true (T) or false (F). Rewrite the false statements to make true statements.

   a._____ Testosterone exerts negative feedback control over the anterior pituitary secretion of LH.

   Rewrite:_____

   b._____ Inhibin exerts negative feedback control over the anterior pituitary secretion of FSH.

   Rewrite: _____

9. What are some of the effects of testosterone on the development of secondary sex characteristics?

   _____

   _____

15.2 FEMALE REPRODUCTIVE SYSTEM (PAGES 294–296)

After you have answered the questions for this section, you should be able to
- Describe the structure and function of the female reproductive system.
- Label a diagram of the external female genitals.

10. Using the alphabetized list of terms and the blanks provided, identify and state a function for the human female reproductive structures and urinary structures shown in the following diagram.

cervix   ovary   oviduct   urethra   urinary bladder   uterus   vagina

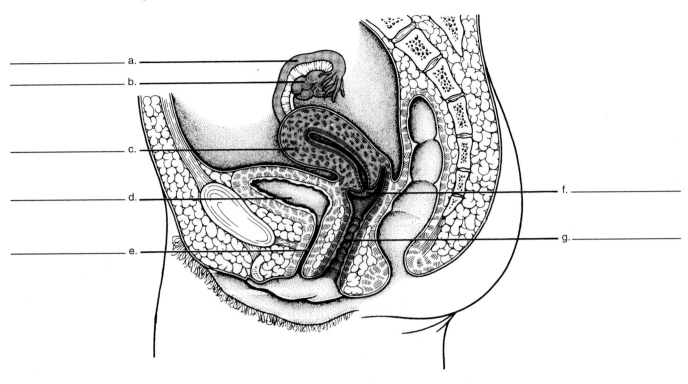

Structure	Function
a. _____	_____
b. _____	_____
c. _____	_____
d. _____	_____
e. _____	_____
f. _____	_____
g. _____	_____

11. When sperm enter the female reproductive tract, they are deposited into the a._____. From there, they pass through the b._____ of the uterus. They swim up through the c._____ until they reach the egg cell.

12. Using the following alphabetized list of terms, label this diagram of the vulva.

anus
glans clitoris
labia majora
labia minora
mons pubis
urethra
vagina

a. _____
b. _____
c. _____
d. _____
e. _____
f. _____
g. _____

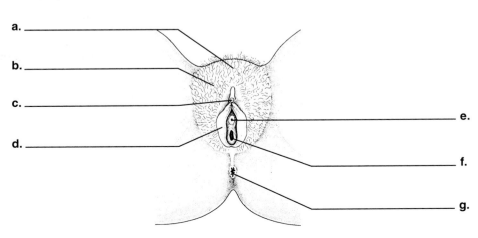

After you have answered the questions for this section, you should be able to
- Describe the ovarian and uterine cycles.
- Discuss hormonal regulation in the female, including feedback control.
- Name the actions of estrogen and progesterone, including the influence on secondary sex characteristics.

13. Each a._____ in the ovary contains an oocyte. A secondary follicle develops into a(n) b._____ follicle. c._____ is the release of the secondary oocyte (egg) from the ovary. Following ovulation, a follicle becomes a(n) d._____.

14. Fill in the following table to indicate the events in the ovarian and uterine cycles (simplified, and assuming a 28-day cycle).

Ovarian Cycle	Events	Uterine Cycle	Events
Follicular phase—days 1–13	a. _____ Follicle maturation occurs. Estrogen secretion is prominent.	b. _____—days 1–5 d. _____—days 6–13	c. _____ e. _____
Ovulation—day 14	LH spike occurs.		
Luteal phase—days 15–28	LH secretion continues. Corpus luteum forms. h. _____	f. _____—days 15–28	g. _____ _____

15. Match the definitions to these terms:

    estrogen   FSH   LH   progesterone

    a. _____ gonadotropic hormones

    b. _____ female sex hormones

    c. _____ causes the endometrium to thicken

    d. _____ a drop in levels is responsible for menses

16. What are some of the effects of estrogen on the development of secondary sex characteristics?

    _____

    _____

17. Indicate whether the following statements are true (T) or false (F). Rewrite the false statements to make true statements.

    a. _____ Implantation occurs as soon as fertilization occurs. Rewrite: _____

    _____

    b. _____ HCG prevents degeneration of the corpus luteum. Rewrite: _____

    _____

    c. _____ During pregnancy, ovulation continues because estrogen and progesterone are still present. Rewrite:

    _____

15.4 CONTROL OF REPRODUCTION (PAGES 301–304)

After you have answered the questions for this section, you should be able to
- Categorize birth control measures by the criteria used in the text.
- List the causes of infertility and the various assisted reproductive technologies.

18. Following are two groups of birth control measures. Rank the members of each group from the most effective (1) to the least effective (4).

A
a. _____ coitus interruptus
b. _____ spermicidal jelly/cream
c. _____ condom + spermicide
d. _____ natural family planning

B
e. _____ vasectomy
f. _____ douche
g. _____ diaphragm + spermicide
h. _____ IUD

19. The most common causes of infertility in males is a._____ caused by
b._____.

20. The most common cause of infertility in females is a._____. Two other causes are
b._____ and c._____.

21. In which assisted reproductive technologies is the egg fertilized in laboratory glassware? _____
_____

DEFINITIONS WORDSEARCH

Review key terms by using the following alphabetized list of terms to fill in the blanks. Then complete the wordsearch.

```
E T A T S O R P A P T E S T R      cervix
N B V C X X S A M U T O R C S      estrogen
T E S T R O G E N T X V V V I      fimbria
L K J H T R E W Q E C A V B T      menopause
P O I C E R V I X R F R C A S      ovary
B E S T A V W S X U H Y Z S E      Pap test
V F I M B R I A S S A D F B T      prostate
G Y E F S M E N O P A U S E L      scrotum
                                   testis
                                   uterus
```

a. _____ Narrow base of the uterus leading to the vagina.
b. _____ Female sex hormone responsible for secondary sex characteristics.
c. _____ Muscular organ in which the fetus develops.
d. _____ Pouch of skin that encloses testes.
e. _____ Fingerlike extension of oviduct.
f. _____ Termination of menstrual cycle in older women.
g. _____ Organ that produces sperm.
h. _____ Organ that produces eggs.
i. _____ Doughnut-shaped gland around male urethra.
j. _____ Clinical test to detect cervical cancer.

Do not refer to the text when taking this test.

____ 1. The vas deferens
   a. becomes erect.
   b. carries sperm.
   c. is surrounded by the prostate gland.
   d. runs through bulbourethral glands.

____ 2. The prostate gland
   a. is removed when a vasectomy is performed.
   b. surrounds the upper portion of the urethra.
   c. receives urine from the bladder.
   d. almost never becomes cancerous.

____ 3. Which gland or organ secretes hormones?
   a. seminal vesicles
   b. prostate gland
   c. bulbourethral gland
   d. testes

____ 4. FSH
   a. stimulates sperm production in males.
   b. stimulates development of the follicle in females.
   c. is produced by the anterior pituitary.
   d. All of these are correct.

____ 5. Gonadotropic hormones are produced by the
   a. testes.
   b. ovaries.
   c. anterior pituitary.
   d. uterus.

____ 6. Which hormone stimulates the production of testosterone?
   a. LH
   b. FSH
   c. estrogen
   d. inhibin

____ 7. Which hormone regulates the production of testosterone?
   a. LH
   b. FSH
   c. estrogen
   d. inhibin

____ 8. The urethra is part of the reproductive tract in
   a. the female.
   b. the male.
   c. both the male and female.
   d. invertebrates.

____ 9. The endometrium
   a. lines the vagina.
   b. receives the developing embryo for implantation.
   c. produces estrogen.
   d. None of these is correct.

____ 10. The uterus
   a. is connected to both the oviducts and the vagina.
   b. is not an endocrine gland.
   c. contributes to the development of the placenta.
   d. All of these are correct.

____ 11. Which structure is present after ovulation?
   a. primary follicle
   b. secondary follicle
   c. vesicular follicle
   d. corpus luteum

____ 12. Ovulation occurs
   a. due to hormonal changes.
   b. always on day 14.
   c. in postmenopausal women.
   d. as a result of intercourse.

____ 13. Which of these secretes hormones involved in the ovarian cycle?
   a. hypothalamus
   b. anterior pituitary gland
   c. ovary
   d. All of these are correct.

____ 14. FSH stimulates the
   a. release of an egg cell from the follicle.
   b. development of a follicle.
   c. development of the endometrium.
   d. development of the corpus luteum.

____ 15. Secretions from which of the following structures are required before implantation can occur?
   a. ovarian follicle
   b. pituitary gland
   c. corpus luteum
   d. All of these are correct.

____ 16. Human chorionic gonadotropin (HCG) is different from other gonadotropic hormones because it
   a. is produced by the hypothalamus.
   b. is not produced by a female endocrine gland.
   c. does not stimulate any tissue in the body.
   d. does not enter the bloodstream.

____ 17. Pregnancy is present
   a. when an egg develops in the corpus luteum.
   b. when ovulation occurs successfully.
   c. following fertilization and implantation.
   d. during the follicular phase only.

____ 18. Menstruation begins in response to
   a. increasing estrogen levels.
   b. decreasing progesterone levels.
   c. hormones from the anterior pituitary.
   d. secretion of FSH.

_____ 19. What do all the birth control methods have in common?
   a. They all use some device.
   b. They all interrupt intercourse.
   c. They are all terribly expensive and uncomfortable.
   d. None of these is correct.

_____ 20. A vasectomy
   a. prevents the egg from reaching the oviduct.
   b. prevents sperm from reaching seminal fluid.
   c. prevents release of seminal fluid.
   d. inhibits sperm production.

_____ 21. Which of these means of birth control prevents implantation?
   a. diaphragm
   b. IUD
   c. cervical cap
   d. vaginal sponge

_____ 22. In vitro fertilization occurs in
   a. the vagina.
   b. a surrogate mother.
   c. laboratory glassware.
   d. the uterus

THOUGHT QUESTIONS

Answer in complete sentences.

23. How do the parts of a sperm assist its function?

24. Why do you expect to find sex hormones from the ovaries in pregnant women but not in menopausal women?

Test Results: _____ number correct ÷ 24 = _____ × 100 = _____%

STUDY QUESTIONS

1. a. urinary bladder; stores urine **b.** vas deferens; conducts and stores sperm **c.** prostate gland; contributes to semen **d.** urethra; conducts both urine and sperm **e.** penis; organ of sexual intercourse **f.** seminal vesicles; contribute to semen **g.** bulbourethral gland; contributes nutrients and fluid to semen **h.** epididymis; stores sperm as they mature **i.** testis; production of sperm and male sex hormones **2. a.** IC **b.** ST **c.** ST **d.** IC **3. a.** epididymis **b.** ejaculatory duct **c.** urethra **4. a.** seminal vesicles **b.** prostate gland **c.** bulbourethral glands **5.** to nourish sperm cells, to increase the motility of sperm cells. **6. a.** spermatogenesis **b.** seminiferous **7. a.** head **b.** middle piece **c.** tail **d.** acrosome **e.** middle piece **8. a.** T **b.** T **9.** Testosterone deepens the voice, promotes the development of muscles, body and facial hair; increases secretions from oil glands; and promotes the development of the sex organs. **10. a.** oviduct; conduction of egg **b.** ovary; production of eggs and sex hormones **c.** uterus; houses developing fetus **d.** urinary bladder; storage of urine **e.** urethra; conduction of urine **f.** cervix; opening of uterus **g.** vagina; receives penis during sexual intercourse and serves as birth canal **11. a.** vagina **b.** cervix **c.** oviduct **12. a.** mons pubis **b.** labia majora **c.** glans clitoris **d.** labia minora **e.** urethra **f.** vagina **g.** anus **13. a.** follicle **b.** vesicular **c.** Ovulation **d.** corpus luteum **14. a.** FSH secretion begins **b.** menstruation **c.** endometrium breaks down **d.** proliferative phase **e.** endometrium rebuilds **f.** secretory phase **g.** endometrium thickens, and glands are secretory **h.** progesterone secretion is prominent **15. a.** FSH and LH **b.** progesterone and estrogen **c.** estrogen **d.** progesterone **16.** Estrogen promotes the deposition of body fat, the maturation and maintenance of the sex organs, and breast development. **17. a.** F, Implantation occurs several days after fertilization. **b.** T **c.** F, During pregnancy, ovulation discontinues because estrogen and progesterone secreted by the corpus luteum and the placenta exert feedback control over the hypothalamus and the anterior pituitary. **18. a.** 3 **b.** 2 **c.** 1 **d.** 4 **e.** 1 **f.** 4 **g.** 3 **h.** 2 **19. a.** low sperm count or large proportion of abnormal sperm **b.** sedentary lifestyle **20. a.** body weight **b.** blocked oviducts **c.** endometriosis **21.** IVF and GIFT

DEFINITIONS WORDSEARCH

```
E T A T S O R P A P T E S T
                M U T O R C S
    E S T R O G E N T   V     I
                    E   A     T
        C E R V I X R   R     S
                    U   Y     E
    F I M B R I A   S         T
        M E N O P A U S E
```

a. cervix **b.** estrogen **c.** uterus **d.** scrotum **e.** fimbria **f.** menopause **g.** testis **h.** ovary **i.** prostate **j.** Pap test

CHAPTER TEST

1. b **2.** b **3.** d **4.** d **5.** c **6.** a **7.** d **8.** b **9.** b **10.** d **11.** d **12.** a **13.** d **14.** b **15.** d **16.** b **17.** c **18.** b **19.** d **20.** b **21.** b **22.** c **23.** A sperm functions to fertilize an egg, and its various parts are specialized; for example, the head is capped by the acrosome, which releases enzymes that allow the sperm to penetrate the egg; the head contains 23 chromosomes; the middle piece contains mitochondria that provide energy; and the tail is a flagellum that allows the sperm cell to swim. **24.** The ovaries do not secrete female sex hormones in pregnant women or in menopausal women. During pregnancy, however, ovarian hormones are replaced by hormones secreted by the placenta. The placental hormones help to maintain the uterine lining and, thus, the pregnancy.

16

DEVELOPMENT AND AGING

Few topics in biology capture the attention of everyone the way human development is able to. We marvel at how one microscopic cell can divide to produce an adult, much less contain all the information to orchestrate each developmental step along the way. In this chapter, you will learn about the stages of development from the embryo to the fetus and about the birthing process. Development from infancy to old age is also discussed.

Make a summary chart of human development, divided into embryonic and fetal development. Include details of cleavage, morula, blastula, and gastrula stages (pp. 311–313 and 315–319).

Understanding fetal circulation requires memorizing blood vessels (see Fig. 16.7). Construct a flow diagram indicating the path of circulation from the mother to the fetus, throughout the fetus, and back to the mother.

Pay attention to how the process of aging can be delayed through proper nutrition and exercise (pp. 327–328), and you can benefit from a healthy, long life.

TIP: Be sure to visit the Online Learning Center that accompanies *Human Biology* 9/e. It has practice quizzes, interactive activities, labeling exercises, art quizzes, animations, flash cards, and much more. http://www.mhhe.com/maderhuman9

Study the text section by section. Answer the study questions so that you can fulfill the learning objectives for each section.

16.1 FERTILIZATION (PAGE 310)

After you have answered the questions for this section, you should be able to
• Describe the process of fertilization, including where and how it occurs.

Fill in the blanks.

1. Fertilization begins when sperm make contact with the a._____, they release enzymes from the b._____, and then one sperm can enter the egg. When the sperm nucleus fuses with the egg nucleus, c._____ is complete.

16.2 DEVELOPMENT BEFORE BIRTH (PAGES 311–321)

After you have answered the questions for this section, you should be able to
• State the four processes of development.
• Name the extraembryonic membranes, and state the function of each.
• Describe the structure and function of the placenta.
• Describe the weekly events of embryonic development and the monthly events of fetal development.
• Trace the path of blood in the fetus.
• Compare the development of gonads in female and male fetuses.

2. Development consists of a progression of events in which (a) the number of cells increases, (b) the size increases, (c) the form and shape of the individual emerges, and (d) the cells become specialized into various organs. What are the terms used to describe these processes?

a._____ b._____

c._____ d._____

3. Using the alphabetized list of terms, label the extraembryonic membranes of the human embryo.

allantois
amnion
chorion
embryo
placenta
yolk sac

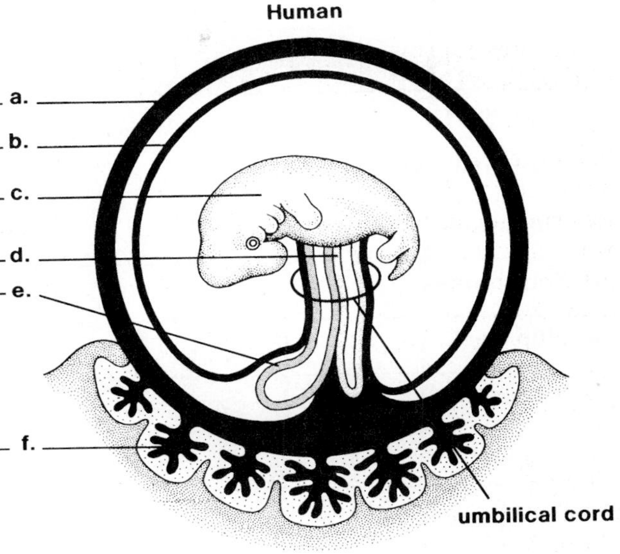

Human

_____ a. _____

_____ b. _____

_____ c. _____

_____ d. _____

_____ e. _____

_____ f. _____

umbilical cord

4. Complete the following table.

Membrane	Function in Human
Chorion	a.
Amnion	b.
Allantois	c.
Yolk sac	d.

5. The process of development. Right after fertilization, the zygote undergoes a series of divisions. The solid mass of cells is called a(n) a._____. Next, an interior cavity forms inside the ball of cells, forming a stage called a(n) b._____ with a(n) c._____ off to one side. It is at this stage that implantation occurs. Next, the inner cell mass develops into a(n) d._____.

6. To describe human embryonic development, complete the following table by writing in the events that occur at times indicated.

   1 all internal organs formed; limbs and digits well formed; recognizable as human, although still small
   2 fertilization; cell division begins
   3 limb buds begin; heart is beating; embryo has a tail
   4 implantation; embryo has tissues; first two extraembryonic membranes
   5 fingers and toes are present; cartilaginous skeleton
   6 nervous system begins; heart development begins
   7 head enlarges; sense organs prominent

Time	Events
a. First week	
b. Second week	
c. Third week	
d. Fourth week	
e. Fifth week	
f. Sixth week	
g. Two months	

7. The placenta has a(n) a._____ side derived from the extraembryonic membrane, the

   b._____, and a(n) c._____ side derived from d._____ tissue.

8. Wastes pass from the fetal side, to the a._____ side and b._____ pass from

   the maternal side to the fetal side.

9. Using the alphabetized list of terms, label the following diagram of fetal circulation.

   arterial duct   oval opening   placenta   umbilical arteries   umbilical vein   venous duct

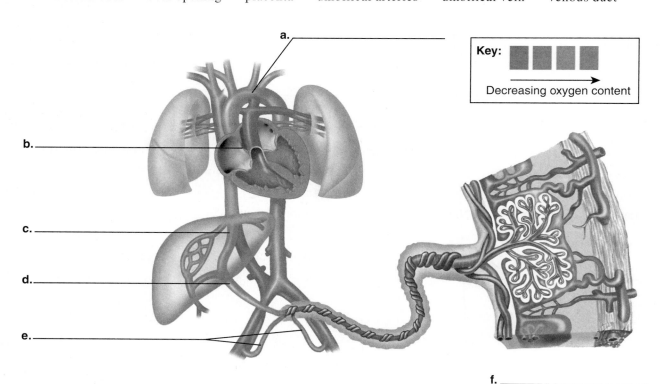

Key:

Decreasing oxygen content

a._____

b._____

c._____

d._____

e._____

f. _____

10. Indicate whether these statements about fetal development (months 3–9) are true (T) or false (F).
    a. _____ It is possible to distinguish sex.
    b. _____ All of the cartilage is replaced by bone.
    c. _____ Limb buds are still present.
    d. _____ Fingernails and eyelashes appear.

11. To describe human fetal development, complete the following table by writing in the events that occur at the times indicated.
    1 possible to distinguish sex; bony skeleton visible; head prominent
    2 skin wrinkled and red, protective cheesy coating; eyes open
    3 ready for birth

Time	Events
a. Third-fourth month	
b. Fifth-seventh month	
c. Eighth-ninth month	

12. It is the genes on the ᵃ·_____ chromosome that determine whether the embryo will become male or female. Gonads begin developing internally in the ᵇ·_____ week of gestation. At six weeks, a small bud appears between the legs of the embryo. It will develop into a(n) ᶜ·_____ if it is a female and into a(n) ᵈ·_____ if it is a male.

16.3 PREGNANCY AND BIRTH (PAGES 324–325)

After you have answered the questions for this section, you should be able to
• Describe the three stages of parturition (birth).

13. What events are associated with the three stages of parturition?
    a. first stage: _____
    b. second stage: _____
    c. third stage: _____

14. True labor is marked by contractions that occur regularly every ᵃ·_____ minutes. The cervix must first undergo ᵇ·_____ before it dilates. The incision made to prevent tearing during childbirth is called a(n) ᶜ·_____.

16.4 DEVELOPMENT AFTER BIRTH (PAGES 326–328)

After you have answered the questions for this section, you should be able to
• Describe the different stages of human development.
• Describe underlying causes of the aging process.

15. Three theories of aging. Some believe that aging is ᵃ·_____ in origin, meaning that our ᵇ·_____ inheritance causes us to age. Others maintain that ᶜ·_____ processes are involved; for example, the hormonal system and the immune system decrease in efficiency as we age. Still others believe that ᵈ·_____ factors influence aging more than we realize; for example, diet and exercise keep us healthy despite added years.

16. Aging affects body systems. Certain systems maintain the body; in regard to the cardiovascular system, ᵃ·_____ disease may be associated with ᵇ·_____ blood pressure, and reduced blood flow to the ᶜ·_____ may result in less efficiency at filtering wastes. In regard to those systems that

integrate and coordinate the body, actually few ^{d.}_____ are lost from the brain, and the elderly can learn new material; loss of skeletal mass and osteoporosis can be controlled by ^{e.}_____. In regard to the reproductive system, there is a reduced level of ^{f.}_____ in both males and females, although males produce sperm until death. Young people should be aware that now is the time to begin the health habits that increase the life span.

DEFINITIONS WORDSEARCH

Review key terms by completing this wordsearch using the following alphabetized list of terms:

```
D F E R G T H Y J U J Z P        amnion
C G E R O N T O L O G Y O        chorion
O U Y T R F G H A D F G A        embryo
L D S W E F R T N E R O M        episiotomy
A U M B I L I C U S D T N        gerontology
L I O U K Y H R G S S E I        lanugo
U M N B J O O P O T Y H O        morula
R D P A R T U R I T I O N        parturition
O R E I Y T H R D C W S E        zygote
M H O E P I S I O T O M Y
V N I O Y R B M E A Q W S
```

a. _____   Study of aging.

b. _____   Extraembryonic membrane that develops into placenta.

c. _____   Extraembryonic membrane that contains protective fluid.

d. _____   Spherical mass of cells resulting from cleavage during animal development.

e. _____   Diploid cell formed after union of two gametes.

f. _____   Fine down that covers the fetus.

g. _____   Process of giving birth.

h. _____   Incision to enlarge vaginal orifices while giving birth.

i. _____   Organism during first eight weeks of development.

CHAPTER TEST

OBJECTIVE TEST

Do not refer to the text when taking this test.

_____ 1. During the process of fertilization
  a. one sperm only enters the egg and fuses with its nucleus.
  b. several sperm enter the egg but only one fuses with its nucleus.
  c. many sperm release acrosome enzymes so that one sperm can enter the egg.
  d. Both _a_ and _c_ are correct.

_____ 2. The hormone that is the basis for the pregnancy test is
  a. estrogen.
  b. follicle-stimulating hormone.
  c. human chorionic gonadotropin (HCG).
  d. progesterone.

In questions 3–6, match the descriptions to the extraembryonic membranes.
  a. placenta
  b. umbilical blood vessels
  c. watery sac
  d. first site of red blood cell formation

_____ 3. chorion
_____ 4. amnion
_____ 5. allantois
_____ 6. yolk sac

_____ 7. Transport occurs at the placenta by the
  a. exchange across embryonic blood vessels.
  b. mixing of maternal and embryonic blood.

_____ 8. Which statement is NOT true of chorionic villi?
  a. They are projections from an extraembryonic membrane.
  b. They stay within the fetal side of the placenta.
  c. They extend into the maternal portion of the placenta.
  d. They initially project from the entire chorion during development.

_____ 9. Which blood vessel in the fetus carries the most oxygen?
  a. pulmonary vein
  b. umbilical vein
  c. arterial duct
  d. venous duct

_____ 10. The oval opening is an opening that allows fetal blood to pass from the
  a. right atrium to the left atrium.
  b. left atrium to the right atrium.
  c. right ventricle to the left ventricle.
  d. pulmonary artery into the aorta.

_____ 11. The fetal circulation is modified because
  a. the heart is not beating.
  b. the lungs are not functioning.
  c. nutrients are obtained from the yolk.
  d. All of these are correct.

_____ 12. The zygote begins to undergo cleavage in the
  a. cervix.
  b. ovary.
  c. oviduct.
  d. uterus.

_____ 13. Which of these pairs is mismatched?
  a. cleavage–cell division
  b. morphogenesis–organ systems present
  c. differentiation–specialization of cells
  d. growth–increase in size

_____ 14. Which is NOT an event normally observed by the sixth week of embryonic development?
  a. head is apparent
  b. limb buds developing
  c. ossification almost complete
  d. tail regressing

_____ 15. Select the incorrect pair related to fetal development.
  a. third month–eyes formed, lids fused
  b. fourth month–ossification continues
  c. fifth month–internal organs maturing
  d. sixth month–4 ounces only

For questions 16–18, match the descriptions to the stages of development.
  a. embryonic development
  b. fetal development
  c. both

_____ 16. heartbeat can be heard by physician
_____ 17. internal organs start developing
_____ 18. extraembryonic membranes formed and functioning

_____ 19. The placenta
  a. takes blood to the developing fetus.
  b. allows exchanges of substances between mother's blood and fetal blood.
  c. forms the umbilical cord.
  d. Both _a_ and _b_ are correct.
  e. All of these are correct.

_____ 20. Which of these is a correct statement?
  a. A male develops when the Y chromosome is present and the testes produce androgens.
  b. A female develops when the X chromosome is present and the ovaries produce estrogen.
  c. Females develop by default—that is, when no Y chromosome is present and no androgens are being produced.
  d. Both _a_ and _b_ are correct.
  e. Both _a_ and _c_ are correct.

_____ 21. A(n) _____ is clearly recognizable as a human being
- a. embryo
- b. fetus
- c. neurula
- d. All of these are correct.

_____ 22. Which system is the first to be visually evident?
- a. nervous
- b. respiratory
- c. digestive
- d. skeletal

_____ 23. The fontanels close after birth generally by _____ months.
- a. 5–8
- b. 9–13
- c. 12–15
- d. 16–24

_____ 24. During which stage of parturition is the baby born?
- a. first
- b. second
- c. third
- d. fourth

_____ 25. Which hormone has no effect on the breasts?
- a. oxytocin
- b. prolactin
- c. estrogen
- d. aldosterone

THOUGHT QUESTIONS

Use the space provided to answer these questions in complete sentences.

26. Why is it important for the extraembryonic membranes to develop early during human development?

27. Distinguish between morphogenesis and differentiation.

Test Results: _____ number correct ÷ 27 = _____ × 100 = _____%

STUDY QUESTIONS

1. **a.** zona pellucida **b.** acrosome **c.** fertilization
2. **a.** cleavage **b.** growth **c.** morphogenesis **d.** differentiation 3. **a.** chorion **b.** amnion **c.** embryo **d.** allantois **e.** yolk sac **f.** placenta 4. **a.** nutrient and gas exchange; becomes placenta **b.** fluid environment; cushioning, protection **c.** becomes umbilical cord for gas and nutrient exchange **d.** first site of red blood cell formation 5. **a.** morula **b.** blastocyst **c.** inner cell mass **d.** embryo 6. Write in the events following these numbers. **a.** 2 **b.** 4 **c.** 6 **d.** 3 **e.** 7 **f.** 5 **g.** 1 7. **a.** fetal **b.** chorion **c.** maternal **d.** uterine 8. **a.** maternal **b.** nutrients 9. **a.** arterial duct **b.** oval opening **c.** venous duct **d.** umbilical vein **e.** umbilical arteries **f.** placenta 10. **a.** T **b.** F **c.** F **d.** T 11. Write in the events following these numbers. **a.** 1 **b.** 2 **c.** 3 12. **a.** Y **b.** 7th **c.** clitoris **d.** penis 13. **a.** dilation of cervix **b.** mother pushes as baby moves down and out of birth canal **c.** afterbirth is expelled 14. **a.** 15–20 **b.** stretching **c.** episiotomy 15. **a.** genetic **b.** genetic **c.** whole-body **d.** extrinsic 16. **a.** cardiovascular **b.** high **c.** kidneys **d.** neurons **e.** exercise **f.** hormones

DEFINITIONS WORDSEARCH

```
                         Z
        G E R O N T O L O G Y
                     A     G A
                     N     O M
    A              C U     T N
    L              H G     E I
    U              O O     O
    R    P A R T U R I T I O N
    O         I
    M    O E P I S I O T O M Y
    N    O Y R B M E
```

a. gerontology **b.** chorion **c.** amnion **d.** morula
e. zygote **f.** lanugo **g.** parturition **h.** episiotomy
i. embryo

CHAPTER TEST

1. d 2. c 3. a 4. c 5. b 6. d 7. a 8. b 9. b
10. a 11. b 12. c 13. b 14. c 15. d 16. b
17. a 18. c 19. b 20. e 21. b 22. a 23. d
24. b 25. d 26. The placenta, which supplies the oxygen and nutrient needs of the embryo and fetus, is derived from the chorion, an extraembryonic membrane. 27. Morphogenesis occurs as cells move and the embryo takes shape; differentiation is apparent when a cell has a particular structure and function—muscle cells look and act differently than nerve cells, for example.

17

CELL DIVISION AND THE HUMAN LIFE CYCLE

STUDY TIPS

This chapter examines whole chromosomes and how they are passed from a parent cell to daughter cells through mitosis (pp. 337–339) and from one generation to the next through meiosis (pp. 340–343). Take a sheet of poster paper and prepare a chart contrasting the stages of mitosis with those of meiosis. Keep in mind that mitosis produces cells for growth or repair, while meiosis produces the gametes that participate in reproduction. Meiosis also has crossing-over (p. 341), a means by which new genetic combinations are possible.

TIP: Be sure to visit the Online Learning Center that accompanies *Human Biology* 9/e. It has practice quizzes, interactive activities, labeling exercises, art quizzes, animations, flash cards, and much more. http://www.mhhe.com/maderhuman9

STUDY QUESTIONS

Study the text section by section. Answer the study questions so that you can fulfill the learning objectives for each section.

17.1 CELL INCREASE AND DECREASE (PAGES 334–335)

After you have answered the questions for this section, you should be able to
- Identify the causes of cell increase and cell decrease.
- Identify the stages of the cell cycle.
- Discuss control of the cell cycle.
- Identify the significance of apoptosis.

1. Label the following diagram of the cell cycle.

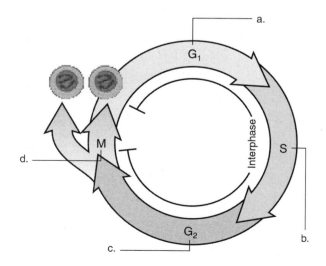

What do these letters stand for?

e. G_1 _____

f. S _____

g. G_2 _____

h. M _____

2. Match a–d in question 1 to these statements.

_____ Mitosis and cytokinesis occur.

_____ DNA replication occurs as chromosomes duplicate.

_____ Growth occurs as cell prepares to divide.

_____ Growth occurs as organelles double.

3. The internal signal a._____ increases and decreases during the cell cycle, while b._____

attempts to repair damaged DNA.

4. Place an **X** on the line beside each statement that is true of apoptosis.

_____ a. Cells are destroyed during apoptosis by external enzymes.

_____ b. Cells are destroyed during apoptosis by internal enzymes.

_____ c. At the end of apoptosis, the cell looks like it did before.

_____ d. At the end of apoptosis, the cell is fragmented.

17.2 MITOSIS (PAGES 337–339)

After you have answered the questions for this section, you should be able to
- Draw and describe the stages of mitosis.

5. Complete the following diagrams to show the arrangement and movement of chromosomes during animal cell mitosis. Briefly describe the events of each phase on the lines provided below.

a. Prophase: _____

b. Metaphase: _____

c. Anaphase: _____

d. Telophase: _____

Prophase

Metaphase

Anaphase

Telophase

After you have answered the questions for this section, you should be able to
- Draw and explain an overview of meiosis.
- Describe the stages and importance of meiosis.

In questions 6–8, fill in the blanks.

6. The nuclear division that reduces the chromosome number from the a._____ number to the

   b._____ number is called meiosis.

7. In a life cycle, the zygote always has the _____ number of chromosomes.

8. A pair of chromosomes having the same length and centromere position are called _____.

9. Indicate whether these statements regarding the role of meiosis are true (T) or false (F):
   a. _____ In animals, meiosis forms gametes that fuse to form a zygote.
   b. _____ Meiosis forms haploid cells in the life cycle of animals.
   c. _____ Meiosis produces four diploid cells over two divisions.

10. Label the following summary diagram of meiosis, using the following alphabetized list of terms.

    centromere
    diploid
    DNA replication
    haploid
    meiosis I
    meiosis II
    nucleolus
    sister chromatids
    synapsis

    Why is it correct to symbolize meiosis as 2n → n?

    j. _____

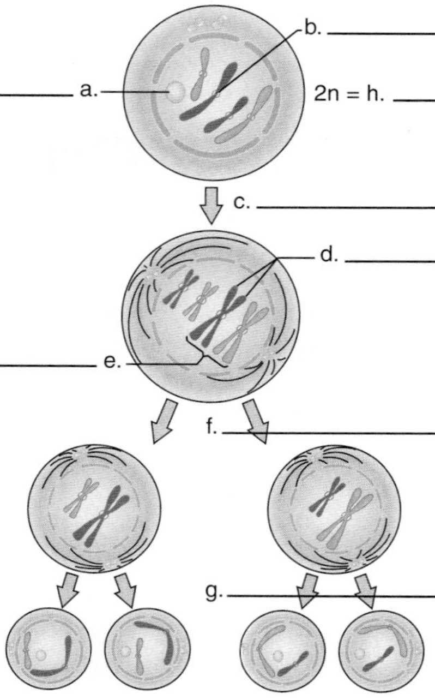

    _____

    a. _____
    b. _____
    2n = h. _____
    c. _____
    d. _____
    e. _____
    f. _____
    g. _____
    n = i. _____

11. a. What is the diploid number of chromosomes in the diagram in question 10? _____
    b. Which structures separate during meiosis I? _____
    c. Which structures separate during meiosis II? _____
    d. Does chromosome duplication occur between meiosis I and meiosis II? _____
    e. Why or why not? _____

12. Match the following terms with the appropriate description:
    crossing-over   genetic variation   synapsis
    a. _____ Homologues line up side by side.
    b. _____ Nonsister chromatids exchange genetic material.
    c. _____ The arrangement of genetic material is new due to crossing-over.

13. Using ink for one duplicated chromosome and pencil for the other, color these homologues before and after crossing-over has occurred.

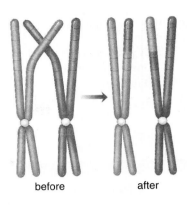

before   after

14. Label and complete the following diagrams to show the arrangement and movement of chromosomes during meiosis I and meiosis II. (The diagram for meiosis II pertains to only one daughter cell from meiosis I.)

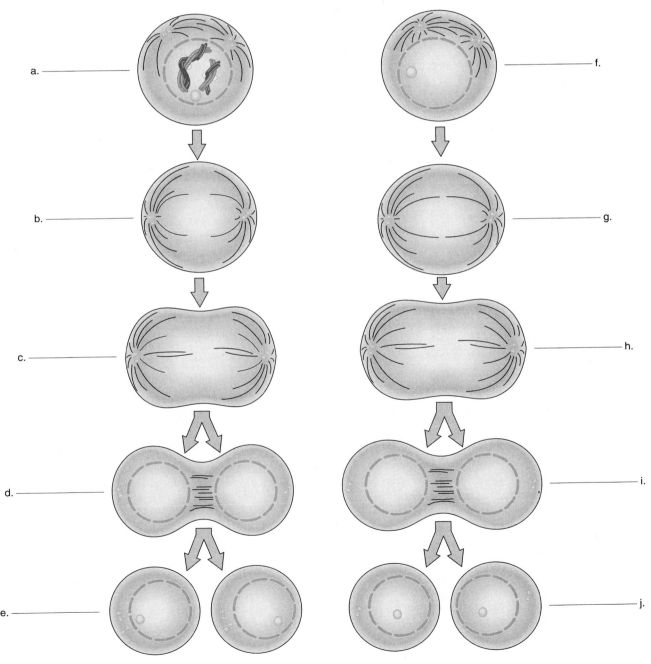

a.

b.

c.

d.

e.

f.

g.

h.

i.

j.

15. Show how genetic recombination occurs as a result of sexual reproduction by matching the following statements:
    1. Zygote carries a unique combination of chromosomes and genes.
    2. Gametes carry different combinations of chromosomes.
    3. Daughter chromosomes carry different combinations of genes.
    a. _____ Homologues separate independently.
    b. _____ Crossing-over occurs.
    c. _____ Gametes fuse during fertilization.

17.4 COMPARISON OF MEIOSIS WITH MITOSIS (PAGES 344–345)

After you have answered the questions for this section, you should be able to
• Understand the differences between the phases of meiosis and mitosis.

16. Complete the table by writing *yes* or *no* to distinguish meiosis from mitosis.

		Meiosis	Mitosis
Completed after one division	a.		
Requires two successive divisions	b.		
During anaphase, sister chromatids separate	c.		
During anaphase I, homologous chromosomes separate	d.		
Results in daughter cells with the diploid number of chromosomes	e.		
Results in daughter cells with the haploid number of chromosomes	f.		
Unique to the somatic (body) cells	g.		
Unique to the sex organs	h.		

17.5 THE HUMAN LIFE CYCLE (PAGES 346–347)

After you have answered the questions for this section, you should be able to
• Describe the roles of mitosis and meiosis in the human life cycle.
• Define spermatogenesis and oogenesis.

17. Indicate whether these statements are true (T) or false (F).
    a. _____ Meiosis in human males is a part of spermatogenesis.
    b. _____ Mitosis in human females is a part of oogenesis.
    c. _____ Oogenesis occurs in the testis.
    d. _____ A zygote undergoes mitosis during the development of the embryo.
    e. _____ Oogenesis produces four functional egg cells from one cell.
18. State whether the following processes occur in males (M), females (F), or both (B).
    a. _____ gamete formation
    b. _____ spermatogenesis
    c. _____ oogenesis
    d. _____ polar body formation

1. Make playing cards with the following words or phrases on them:

    1. *nucleolus*
    2. *chromosomes (chromatids held by centromere)*
    3. *centrosome with centrioles*
    4. *nucleus (nuclear envelope)*
    5. *spindle fibers*
    6. *aster*
    7. *furrowing*
    a. From these, select those that have to do with the formation and structure of the spindle.
    b. From the cards remaining, select those that name structures that disappear during mitosis.
    c. Select the cards that have to do with cytokinesis.

2. Make cards with the following phrases on them:
    1. *chromatids separate*
    2. *distinct chromosomes within daughter nuclei*
    3. *chromosomes arranged at the equator*
    4. *chromosomes are distinct; spindle appears; nucleolus disappears and nuclear envelope fragments*
    5. *homologous chromosomes separate*
    a. Arrange the cards to describe the events of mitosis or meiosis II.
    b. Arrange the cards to describe the events of meiosis I.

3. Make three cards marked as follows:
    16
    8 (diploid number)
    4 (haploid number)
    a. Pick the card that tells how many chromosomes are in the parent nucleus before division.
    b. Pick the card that tells how many chromosomes each daughter nucleus will have after mitosis.
    c. Pick the card that tells how many chromosomes are in the parent nucleus during prophase of mitosis or prophase I of meiosis.
    d. Pick the card that tells how many chromatids are in the parent nucleus during prophase of mitosis or prophase I of meiosis. (For purposes of the game, assume the chromosomes are duplicated.)
    e. Pick the card that tells how many chromosomes are at *both* poles during anaphase of mitosis.
    f. Pick the card that tells how many chromosomes are at *both* poles during anaphase I of meiosis.
    g. Pick the card that tells how many chromosomes are in each daughter nucleus after meiosis I.
    h. Pick the card that tells how many chromosomes are in each daughter nucleus after meiosis II.

DEFINITIONS WORDSEARCH

Review key terms by completing this wordsearch using the following alphabetized list of terms:

```
S E S A H P R E T N I G C V F
A W A S E A D W R F T H E G Y
P E S Y N A P S I S G F N N U
O D I S D E F L G T J Y T P L
P P S L I K J U O G T F R K O
T T O G Y H U J K I H H O Y U
O R T T D I P L O I D S M B N
S P I J I J H S I S O I E M S
I G M O O G E N E S I S R R D
S V Y H F R F R D V C X E F C
S P E R M A T O G E N E S I S
```

apoptosis
centromere
diploid
haploid
interphase
meiosis
mitosis
oogenesis
spermatogenesis
synapsis

a. _____ Cell division resulting in cells identical to the parental cell.

b. _____ Cell division resulting in cells with half as many chromosomes as the parental cell.

c. _____ Process of egg production in females.

d. _____ Process of sperm production in males.

e. _____ Programmed cell death.

f. _____ Region where sister chromatids remain attached.

g. _____ Stage of cell cycle between cell divisions.

h. _____ Pairing of homologous chromosomes during prophase I.

i. _____ Half the usual number of chromosomes.

j. _____ The normal full complement of chromosomes.

CHAPTER TEST

OBJECTIVE TEST

Do not refer to the text when taking this test. In questions 1–4, match the description to the phase.
   a. anaphase
   b. metaphase
   c. prophase
   d. telophase

____ 1. Chromosomes first become visible.
____ 2. Chromatids separate at centromere.
____ 3. Chromosomes are aligned at the equator.
____ 4. Last phase of nuclear division.

____ 5. The diploid chromosome number in an organism is 42. The number of chromosomes in its sex cells is normally
   a. 21.
   b. 42.
   c. 63.
   d. 84.

____ 6. Which description of mitosis is NOT correct?
   a. does not affect the chromosome number
   b. forms four daughter cells
   c. makes diploid nuclei
   d. prophase is the first active phase

____ 7. Which of the following occurs in anaphase of mitosis?
   a. Chromosomes first appear.
   b. Homologous chromosomes separate.
   c. The nuclear envelope breaks down.
   d. Sister chromatids separate.

____ 8. Select the incorrect association.
   a. G_1—cell grows in size
   b. G_2—protein synthesis occurs
   c. mitosis—nuclear division
   d. S—DNA fails to duplicate

_____ 9. The phase of cell division in which the nuclear envelope and nucleolus are disappearing as the spindle fibers are appearing is called
a. anaphase.
b. prophase.
c. telophase.
d. metaphase.

_____10. In animal cells, cytokinesis takes place by
a. membrane fusion.
b. a furrowing process.
c. formation of a cell plate.
d. cytoplasmic contraction.

_____11. Cyclin
a. is a molecule that regulates the cell cycle.
b. binds sister chromatids together.
c. makes up the spindle.
d. All of these are correct.

_____12. If a cell is to divide, DNA replication must occur during
a. prophase.
b. metaphase.
c. anaphase.
d. telophase.
e. interphase.

_____13. If a cell had 18 chromosomes, how many chromosomes would each daughter cell have after mitosis?
a. 9
b. 36
c. 18
d. The number cannot be determined.

_____14. Normal growth and repair of the human body requires
a. mitosis.
b. binary fission.
c. Both _a_ and _b_ are correct.
d. Neither _a_ nor _b_ is correct.

_____15. The cell cycle
a. includes mitosis as an event.
b. includes only the stages G_1, S, and G_2.
c. is under cellular but not under genetic control.
d. involves proteins but not chromosomes.

_____16. When do chromosomes move to opposite poles?
a. prophase
b. metaphase
c. anaphase
d. telophase

In questions 17–23, label each statement with one of the following choices:
a. meiosis I
b. meiosis II

_____17. Synapsis of homologues occurs.
_____18. Separation of homologues occurs.
_____19. Results in one oocyte and two polar bodies in human females.
_____20. Results in four sperm cells in human males.
_____21. Preceded by DNA replication.
_____22. Preceded by interkinesis.
_____23. Crossing-over occurs.

_____24. Which of the following is NOT a valid contrast between mitosis and meiosis?

Mitosis	Meiosis
a. Requires one set of phases	Requires two sets of phases
b. Occurs when somatic (body) cells divide	Occurs during gamete production
c. Results in four daughter nuclei	Results in two daughter nuclei
d. Results in daughter nuclei with diploid number of chromosomes	Results in daughter nuclei with haploid number of chromosomes

_____25. Polar bodies are formed during
a. meiosis.
b. mitosis.
c. oogenesis.
d. spermatogenesis.
e. Both _a_ and _c_ are correct.

_____26. During anaphase of meiosis II,
a. homologues separate.
b. sister chromatids separate.
c. daughter centrioles separate.
d. duplicated chromosomes separate.

_____27. During interkinesis,
a. chromosome duplication occurs.
b. chromosomes consist of two chromatids.
c. meiosis I is complete.
d. Both _b_ and _c_ are correct.

_____28. By the end of meiosis I,
a. crossing-over has occurred.
b. sister chromatids have separated.
c. synapsis of homologues has occurred.
d. each daughter nucleus is genetically identical to the original cell.
e. Both _a_ and _c_ are correct.

Answer in complete sentences.

29. Compare and contrast oogenesis and spermatogenesis.

30. How does sexual reproduction increase genetic variations?

Test Results: _____ number correct ÷ 30 = _____ × 100 = _____ %

ANSWER KEY

STUDY QUESTIONS

1. a. Organelles begin to double in number **b.** Replication of DNA **c.** Sythesis of proteins **d.** Mitosis **e.** Growth **f.** Synthesis **g.** Growth **h.** Mitosis **2.** d, b, c, a **3. a.** cyclin **b.** p 53 **4.** b and, d are true **5. a.** Chromosomes are now distinct; nucleolus is disappearing; centrosomes begin moving apart, and nuclear envelope is fragmenting. **b.** Chromosomes are at the equator. **c.** Daughter chromosomes move to the poles of the spindle. **d.** Daughter cells are forming as nuclear envelopes and nucleoli appear

SOLITAIRE

1. a. 3, 5, 6 **b.** 1, 4 **c.** 7 **2. a.** 4, 3, 1, 2 **b.** 4, 3, 5, 2 **3. a.** 8 **b.** 8 **c.** 8 **d.** 16 **e.** 16 **f.** 8 **g.** 4 **h.** 4

Prophase

Metaphase

Anaphase

Telophase

6. a. diploid (2n) **b.** haploid (n) **7.** diploid (2n) (or full) **8.** homologues or homologous chromosomes **9. a.** T **b.** T **c.** F **10. a.** nucleolus **b.** centromere **c.** DNA replication **d.** sister chromatids **e.** synapsis **f.** meiosis I **g.** meiosis II **h.** diploid **i.** haploid **j.** A diploid cell becomes haploid. The parent cell is diploid and undergoes meiosis, which results in four daughter cells, each of which is haploid **11. a.** 4 **b.** homologues or homologous chromosomes **c.** sister chromatids **d.** no **e.** The chromosomes are already duplicated. **12. a.** synapsis **b.** crossing-over **c.** genetic variation **13.** See Fig. 17.8, page 341, in text. **14. a.** prophase I **b.** metaphase I **c.** anaphase I **d.** telophase I **e.** interkinesis **f.** prophase II **g.** metaphase II **h.** anaphase II **i.** telophase II **j.** daughter cells. See Fig. 17.10, p. 342–343, in text. **15. a.** 2 **b.** 3 **c.** 1 **16. a.** no; yes **b.** yes; no **c.** no; yes **d.** yes; no **e.** no; yes **f.** yes; no **g.** no; yes **h.** yes; no **17. a.** T **b.** F **c.** F **d.** T **e.** F **18. a.** B **b.** M **c.** F **d.** F

DEFINITIONS WORDSEARCH

```
      E S A H P R E T N I   C
A               A           E
P   S Y N A P S I S         N
O       I       L           T
P   S       O       I       R
T   O           I       O   M
O   T   D I P L O I D     M
S   I           S I S O I E M
I   M O O G E N E S I S R
S                         E
S P E R M A T O G E N E S I S
```

a. mitosis **b.** meiosis **c.** oogenesis **d.** spermatogenesis **e.** apoptosis **f.** centromere **g.** interphase **h.** synapsis **i.** haploid **j.** diploid

CHAPTER TEST

1. c **2.** a **3.** b **4.** d **5.** a **6.** b **7.** d **8.** d **9.** b **10.** b
11. a **12.** e **13.** c **14.** a **15.** a **16.** c **17.** a **18.** a
19. b **20.** b **21.** a **22.** b **23.** a **24.** c **25.** e **26.** b
27. d **28.** e **29.** Oogenesis results in one egg. Spermatogenesis results in four sperm. Both undergo meiosis I and meiosis II to produce haploid cells. **30.** Sexual reproduction results in genetic recombinations among offspring due to (1) crossing-over of nonsister chromatids, (2) independent separation of homologues, and (3) fertilization.

18

PATTERNS OF INHERITANCE

The inheritance of traits from our parents and their passage on to our children is a fascinating area of study and the subject of this chapter. Master the terminology *first* (heterozygous vs. homozygous, dominant vs. recessive—see p. 352); then begin to work problems. The first step to problem solving is to write out genotypes and phenotypes (see pp. 352, 354). Then, list possible gametes *before* you try to cross them. It is a common error to try to cross two genotypes between individuals rather than crossing their gametes. Next, use a Punnett square (p. 354) to set up your cross. On some problems it is not necessary to use a Punnett square, but it is good to get into the habit of doing so. Then, work as many problems as your time allows. Know the examples of multifactorial inheritance (p. 359) and multiple alleles (p. 361). Among the more interesting anomalies that occur are those carried on the sex chromosomes (pp. 362). Include them in your study as well.

TIP: Be sure to visit the Online Learning Center that accompanies *Human Biology* 9/e. It has practice quizzes, interactive activities, labeling exercises, art quizzes, animators, flash cards, and much more. http://www.mhhe.com/maderhuman9

STUDY QUESTIONS

Study the text section by section. Answer the study questions so that you can fulfill the learning objectives for each section.

18.1 GENOTYPE AND PHENOTYPE (PAGE 352)

After you have answered the questions for this section, you should be able to
• Distinguish between phenotype and genotype, dominant and recessive, and homozygous and heterozygous.

In question 1, fill in the blank.

1. Letters on homologous chromosomes stand for genes that control a(n) a. _____ , such as hair color.

   Capital letters represent b. _____ alleles, while lowercase letters represent c. _____

   alleles. An individual with a(n) d. _____ of *Aa* is said to be e. _____ and would

   exhibit a dominant f. _____.

18.2 ONE- AND TWO-TRAIT INHERITANCE (PAGES 353–358)

After you have answered the questions for this section, you should be able to
• Solve one-trait autosomal genetics problems using a Punnett square.
• List a variety of simple autosomal dominant and autosomal recessive traits in humans.

2. Give the genotype of an individual who is heterozygous for earlobe attachment (*E* = unattached earlobe,

   e = attached). _____

3. What are the gametes for this person? _____

4. Using this key, E = unattached, e = attached, do Punnett squares for cross 1 and cross 2.

Cross 1
heterozygous × homozygous recessive

Cross 2
heterozygous × heterozygous

    a. What is the phenotypic ratio for cross 1? _____

    b. What are the chances of the recessive phenotype for cross 1? _____

    c. What is the phenotypic ratio for cross 2? _____

    d. What are the chances of the recessive phenotype for cross 2? _____

5. A man with a straight hairline reproduces with a woman with a widow's peak whose father has a straight hairline and whose mother had a widow's peak. The couple produce a child with a straight hairline. What are the genotypes of these individuals involved?
    Use the alleles W (widow's peak) and w (straight hairline).

    a. man with a straight hairline _____ _____

    b. woman with a widow's peak _____ _____

    c. child _____ _____

    d. woman's father _____ _____

In questions 6–7, fill in the blanks.

6. The cross $WwSs \times WwSs$ usually results in a phenotypic ratio close to 9:3:3:1. If W = widow's peak, w = straight hairline, S = short fingers, and s = long fingers, then out of 16 individuals:

9 individuals are expected to have the phenotype a. _____.

3 individuals are expected to have the phenotype b. _____.

3 individuals are expected to have the phenotype c. _____.

1 individual is expected to have the phenotype d. _____.

7. A parent with a widow's peak and short fingers has a child with a straight hairline and long fingers. What is the genotype of the parent? a. _____ If the parent were homozygous dominant for both traits, what would be the phenotype of the child? b. _____

8. a. Complete a Punnett square for the cross in horses $BbTt \times Bbtt$, where:

        B = black
        b = brown
        T = trotter
        t = pacer

    b. What is the phenotypic ratio among offspring? _____ : _____ ;

_____ : _____

18.3 BEYOND SIMPLE INHERITANCE PATTERNS (PAGES 359–361)

After you have answered the questions for this section, you should be able to
- Discuss one trait inherited as a polygenic trait.
- Explain what is meant by multiple alleles, and give an example.
- Explain what is meant by incomplete dominance, and give an example.

9. A woman has a genotype of *AABB* for skin color. She reproduces with a man whose genotype is *aabb*. What are the genotype and phenotype of their offspring?

   a. genotype _____ b. phenotype _____

10. Mrs. Doe and Mrs. Roe had babies at the same hospital. Mrs. Doe took home a girl, Nancy, and Mrs. Roe received a boy, Richard. However, Mrs. Roe was certain she had given birth to a girl and brought suit against the hospital. Blood tests showed that Mr. Roe was type O, Mrs. Roe was type AB, and Mr. and Mrs. Doe were both type B. Nancy was type A, and Richard was type O. List the possible genotypes of each individual, and then indicate the correct parents for each child.

    Genotypes:

    a. Mrs. Doe _____   d. Mrs. Roe _____

    b. Mr. Doe _____   e. Mr. Roe _____

    c. Nancy _____   f. Richard _____

    g. Whose baby is Nancy? _____

    h. Whose baby is Richard? _____

11. Curly and straight hair, when crossed, are an example of _____ dominance because the resulting individuals have the intermediate characteristic of wavy hair.

12. A curly-haired man has children with a wavy-haired woman. What are the genotypes and phenotypes of their children?

    Use the alleles *hh* (curly hair) and *HH* (straight hair).

    Man's genotype ᵃ·_____ Woman's genotype ᵇ·_____
    Use this Punnett square to determine the outcome of this cross.

    c. Genotypes of children _____

    d. Possible phenotypes of children _____

    e. Could any of the children have straight hair? _____

13. A child has sickle-cell disease. What are the genotypes of the parents, who appear to be normal? _____

18.4 SEX-LINKED INHERITANCE (PAGES 362–363)

After you have answered the questions for this section, you should be able to
- Show how sex-linked traits are generally inherited on the X chromosome.
- Discuss one sex-linked disorder.

14. If X^B = normal vision and X^b = color blindness, state the sex and the phenotype of each of these genotypes:

    $X^B X^B$ a. _____

    $X^B X^b$ b. _____

    $X^b X^b$ c. _____

    $X^B Y$ d. _____

    $X^b Y$ e. _____

15. Under what circumstances can a couple have a color-blind daughter? _____

16. Use the Punnett square at the right to show the expected outcome if a color-blind woman reproduces with a man who has normal vision.

   a. What are the chances of a color-blind daughter? _____

   b. What are the chances of a color-blind son? _____

17. A son is color-blind, but his mother and father are not color blind. Give the genotype of all persons involved.

   son _____

   mother _____

   father _____

DEFINITIONS WORDSEARCH

Review key terms by completing this wordsearch, using the following alphabetized list of terms:

```
M E P O L O C U S I C
U F H T H J K L O P D
L G E N O T Y P E U R
T C N F B T F D E S O
I O O X L I N K E D I
P P T I K Y H T G E E
L Y Y E D S R F G T R
E B P U N N E T T J U
F T E L E L L A P O L
```

allele
genotype
locus
multiple
phenotype
Punnett
X-linked

a. _____ Alternative form of a gene on a chromosome.

b. _____ Outward expression of a gene.

c. _____ Particular site where a gene is found on a chromosome.

d. _____ Genetic makeup of an individual.

e. _____ Type of square used to determine genetic outcome.

f. _____ Carried on the X chromosome.

g. _____ More than two alleles for one trait are present in the population. (_____ allele)

CHAPTER TEST

OBJECTIVE TEST

Do not refer to the text when taking this test.

____ 1. If 25% of the offspring of one set of parents show the recessive phenotype, the parents were probably
   a. both homozygous recessive.
   b. both homozygous dominant.
   c. both heterozygous.
   d. one homozygous dominant, one homozygous recessive.

____ 2. Alleles
   a. are alternate forms of a gene.
   b. have the same position on a pair of chromosomes.
   c. affect the same trait.
   d. All of these are correct.

____ 3. Which of these crosses could produce a blue-eyed child? (*B* = brown, *b* = blue) (There may be more than one answer.)
   a. $BB \times bb$
   b. $Bb \times Bb$
   c. $bb \times Bb$
   d. $Bb \times BB$
   e. $Bb \times bb$

Use this information to answer questions 4 and 5: All of the boys born to normal parents are color-blind, while all the daughters are not.

____ 4. The disorder is
   a. X-linked.
   b. dominant.
   c. recessive.
   d. not able to be determined.

_____ 5. The parents are
   a. $X^B X^B \times X^B Y$.
   b. $X^B X^b \times X^B Y$.
   c. $X^B X^B \times X^b Y$.
   d. $X^B X^b \times X^b Y$.

_____ 6. If a man is a carrier of a disease, but a woman is homozygous normal, what are the chances of their child having the disease?
   a. none
   b. 50%
   c. 25%
   d. 1:1

_____ 7. Why is it that two normal parents can have a child with a genetic disorder?
   a. It is a dominant inherited disorder.
   b. It is a recessive inherited disorder.
   c. It results due to an error in gamete formation.
   d. There is no known explanation.

_____ 8. If only one parent is a carrier for a dominant disease, what is the chance a child will have the condition?
   a. 25%
   b. 50%
   c. 75%
   d. no chance

_____ 9. Polygenic inheritance can explain
   a. a range in phenotypes among the offspring.
   b. the occurrence of degrees of dominance.
   c. incomplete dominance.
   d. Both a and c are correct.

_____ 10. Color blindness illustrates
   a. dominance.
   b. multifactorial inheritance.
   c. incomplete dominance.
   d. X-linked inheritance.

_____ 11. Two individuals with medium-brown skin color could have children who are both darker and lighter than they are.
   a. true
   b. false

_____ 12. A female with light skin would be able to have a child with very dark skin if she reproduces with a very dark-skinned male, or a dark child if she reproduced with a light-skinned male.
   a. true
   b. false

_____ 13. Which children could not have parents both with type A blood?
   a. type A
   b. type O
   c. type AB
   d. type B
   e. Both c and d are correct.

_____ 14. Which children could not have a parent with type AB blood?
   a. type A
   b. type B
   c. type AB
   d. type O

_____ 15. Inheritance by multiple alleles is illustrated by the inheritance of
   a. skin color.
   b. blood type.
   c. color blindness.
   d. Both b and c are correct.

_____ 16. There are multiple recessive alleles for Rh⁻, but they are all recessive to Rh⁺.
   a. true
   b. false

Use the following information for questions 17–18: A mother with a genetic disease has genotype of _Aa_. The father, who is normal, has a genotype of _aa_.

_____ 17. This trait is inherited as a _____ disease.
   a. dominant
   b. recessive
   c. incomplete dominant
   d. X-linked

_____ 18. What percentage of their children could have the disease?
   a. none
   b. 25%
   c. 50%
   d. 100%

_____ 19. A woman who is not color-blind but has an allele for color blindness reproduces with a man who has normal color vision. What is the chance they will have a color-blind daughter?
   a. 50%
   b. 25%
   c. 100%
   d. no chance

_____ 20. A color-blind woman reproduces with a man who has normal color vision. Their sons will
   a. be like the father because the trait is X-linked.
   b. be like the mother because the trait is X-linked.
   c. all have normal color vision.

In questions 21–24, match the phenotype with these genotypes.

   a. $X^B X^b$
   b. $X^b Y$
   c. $X^B Y$
   d. $X^B X^B$

_____ 21. male, color-blind
_____ 22. female, carrier for color blindness
_____ 23. male, normal vision
_____ 24. female, normal vision
_____ 25. A girl is color-blind (sex-linked recessive trait).
   a. She received a color-blind allele from her mother.
   b. She received a color-blind allele from her father.
   c. All her sons will be color-blind.
   d. Her father is color-blind.
   e. All of the above are correct.

Use the space provided to answer these questions in complete sentences.

26. To test whether an animal has a homozygous dominant genotype or a heterozygous genotype, it is customary to mate it to the homozygous recessive animal rather than a heterozygote. Why?

27. Males who have a sex-linked genetic disorder most likely inherited the abnormal allele from their mothers. Why?

Test Results: _____ number correct ÷ 27 = _____ × 100 = _____%

ANSWER KEY

STUDY QUESTIONS

1. a. trait **b.** dominant **c.** recessive **d.** genotype **e.** heterozygous **f.** phenotype   **2.** *Ee*   **3.** *E* and *e*   **4. a.** 1:1 **b.** 50% **c.** 3:1 **d.** 25%   **5. a.** *ww* **b.** *Ww* **c.** *ww* **d.** *ww*   **6. a.** widow's peak and short fingers **b.** widow's peak and long fingers **c.** straight hairline and short fingers **d.** straight hairline and long fingers   **7. a.** *WwSs* **b.** widow's peak and short fingers
8. a.

	Bt	bt
BT	BBTt	BbTt
Bt	BBtt	Bbtt
bT	BbTt	bbTt
bt	Bbtt	bbtt

b. 3 black trotters: 3 black pacers; 1 brown pacer: 1 brown trotter   **9. a.** *AaBb* **b.** medium-brown skin color   **10. a.** *iBi* **b.** *iBi* **c.** *iAi* **d.** *iAiB* **e.** *ii* **f.** *ii* **g.** Nancy belongs to the Roes. **h.** Richard belongs to the Does.   **11.** incomplete   **12. a.** *hh* **b.** *hh* **c.** *Hh, hh* **d.** curly or wavy hair **e.** no, because both parents must pass on an *H*   **13.** both are heterozygous for the trait   **14. a.** female with normal vision **b.** female who is a carrier **c.** female who is color-blind **d.** male with normal vision **e.** male who is color-blind   **15.** If the father is color-blind and the mother is a carrier of color blindness.   **16. a.** none **b.** 100%
17. son X^bY, mother X^BX^b, father X^BY

DEFINITIONS WORDSEARCH

```
M   P   L O C U S
U   H
L G E N O T Y P E
T   N
I   O X L I N K E D
P   T
L   Y
E   P U N N E T T
    E L E L L A
```

a. allele **b.** phenotype **c.** locus **d.** genotype **e.** Punnett **f.** X-linked **g.** multiple

CHAPTER TEST

1. c   **2.** d   **3.** b, c, e   **4.** a   **5.** b   **6.** a   **7.** b   **8.** b
9. a   **10.** d   **11.** a   **12.** b   **13.** e   **14.** d   **15.** b
16. a   **17.** a   **18.** c   **19.** d   **20.** b   **21.** b   **22.** a
23. c   **24.** d   **25.** e   **26.** If a heterozygote is mated to a heterozygote, there is a 25% chance for any offspring to be recessive. If a heterozygote is mated to a homozygous recessive, there is a 50% chance for any offspring to be recessive. It is only when the animal is mated to a homozygous recessive that its genotype can be determined.   **27.** Males inherit only a Y sex chromosome from their fathers. The Y chromosome is blank for most sex-linked traits, and therefore, the abnormal allele must be on the X chromosome inherited from the mother.

19

DNA BIOLOGY AND TECHNOLOGY

In this chapter, you will learn about the exciting field of biotechnology. However, you must first understand the basic structure and function of the nucleic acids that form the foundation for this field.

Compare the structure of DNA (pp. 368–369) with that of RNA (p. 370). Remember that in biology, structure relates to function. Learn how the structure of DNA allows it to be the genetic material for the cell and how RNA can be the messenger between the DNA and the cytoplasm of the cell.

When you study the section on gene expression (pp. 371–377), draw a step-by-step diagram illustrating each event from transcription to translation. You may need to review it several times to have a thorough understanding of gene expression. Understand how the cell regulates gene expression (p. 377).

Study the diagrams indicating how recombinant DNA is manufactured. Produce your own diagram if need be. Study the polymerase chain reaction (p. 381), and understand how it helps mass-produce copies of DNA. Be ready to give several examples of how this reaction is employed in biology and other fields.

Make note of types of genetically engineered organisms produced thus far (pp. 383–384).

TIP: Be sure to visit the Online Learning Center that accompanies *Human Biology* 9/e. It has practice quizzes, interactive activities, labeling exercises, art quizzes, animators, flash cards, and much more. http://www.mhhe.com/maderhuman9

STUDY QUESTIONS

Study the text section by section. Answer the study questions so that you can fulfill the learning objectives for each section.

19.1 DNA AND RNA STRUCTURE AND FUNCTION (PAGES 368–370)

After you have answered the questions for this section, you should be able to
- Describe the structure of the DNA double helix.
- Explain what is meant by complementary base pairing.
- Understand how DNA replicates.
- Describe the structure and function of the three types of RNA.
- Describe the differences between the structure of DNA and RNA.

1. Indicate whether these statements concerning DNA structure are true (T) or false (F). Rewrite if false.
   a. _____ DNA is a nucleic acid found in the nucleus. Rewrite: _____

   _____

   b. _____ DNA is a polynucleotide. Rewrite: _____

   _____

   c. _____ Each nucleotide in DNA is composed of the sugar ribose, a phosphate, and a base. Rewrite: ____

   _____

   d. _____ DNA is a double-stranded helix. Rewrite: _____

   _____

2. Examine the "ladder" structure of DNA in the following diagram and answer the questions that follow.

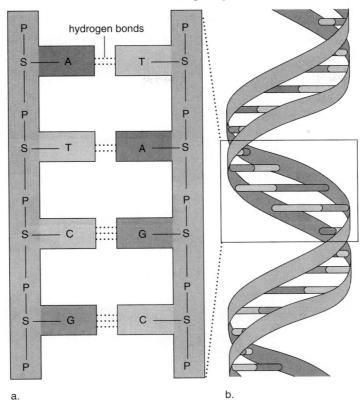

a.   b.

a. Which types of molecules make up the sides of the ladder? _____ and _____

b. Which types of molecules make up the rungs of the ladder? _____

c. Explain complementary base pairing. _____

d. What type of bond joins the bases? _____

e. What do you do to the ladder structure to make a double helix? _____

3. Study the following diagram of replication.

a. The bases in parental DNA are held together by what type of bond (not shown)? _____

b. What happens to these bonds for replication to take place? _____

c. During replication, new nucleotides move into proper position by what methodology? _____

d. Elongation of DNA is catalyzed by an enzyme called _____. When replication is finished there will be two DNA molecules.

e. Each double helix consists of a(n) _____ strand and a(n) _____ strand. Therefore, the process is called _____.

f. Each double helix has _____ (the same, a different) sequence of complementary paired bases.

4. Because genes (DNA) reside in the ᵃ·_____ of the cell and polypeptide synthesis occurs in the ᵇ·_____, they must have a go-between. The most likely molecule to fill this role is ᶜ·_____.

5. Indicate whether these statements about differences between DNA and RNA are true (T) or false (F):
   a. _____ DNA is double stranded; RNA is single stranded.
   b. _____ DNA is a polymer; RNA is a building block of that polymer.
   c. _____ DNA occurs in three forms; RNA occurs in only one form.
   d. _____ The sugar of DNA is ribose, absent in RNA.
   e. _____ Uracil, in RNA, replaces the base thymine, found in DNA.

6. Complete the following table to describe the function of the various types of RNA involved in protein synthesis.

RNA	Function
Ribosomal RNA (rRNA)	a. _____
Messenger RNA (mRNA)	b. _____
Transfer RNA (tRNA)	c. _____

19.2 GENE EXPRESSION (PAGES 371–377)

After you have answered the questions for this section, you should be able to
- Describe the structure and function of proteins.
- Explain what is meant by DNA's triplet code.
- Define the two steps of gene expression—transcription and translation.
- Describe the three steps that take place during translation.
- Understand how the cell controls gene expression.

7. The following diagram represents portions of two different proteins.

   Each circle in the diagram represents a(n) ᵃ·_____. Every protein has its own particular sequence of ᵇ·_____.

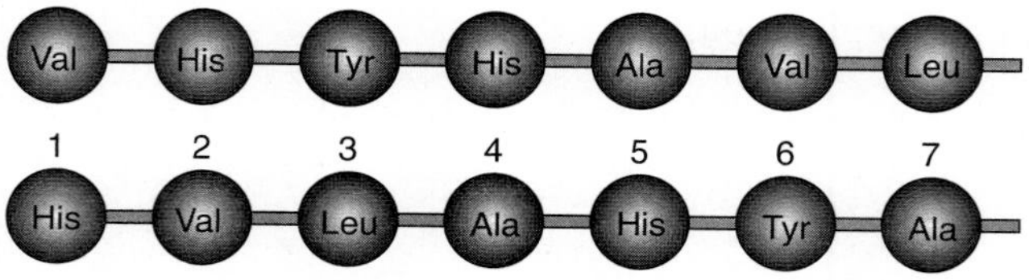

   Would the amino acid at number 3 necessarily be the same in both proteins? ᶜ·_____ At any number would the amino acids necessarily be the same? ᵈ·_____ How is one protein different from another?
   ᵉ·_____

   How do proteins function in metabolic pathways? ᶠ·_____

8. Label the following diagram, which shows the role of DNA, mRNA, and tRNA during gene expression.

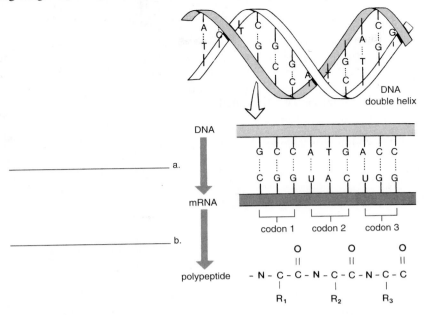

a. _____

b. _____

9. Whereas DNA has a triplet code—that is, every three a._____ stand for a(n) b._____, mRNA contains c._____, each made up of three bases that stand for an amino acid.

10. Complete this paragraph to describe transcription and mRNA processing.

During transcription, an RNA molecule is formed that has a sequence of bases a._____ to a portion of one DNA strand. The bases pair in this manner: A in DNA pairs with b._____, and G pairs with c._____ (and vice versa) in the mRNA being formed. If the sequence of bases in DNA is CGA AGC TCT, then the sequence in mRNA is d. _____
Why is there a space between every three bases? e._____

_____

Which of these—exons or introns—is spliced out when primary mRNA is processed? f. _____
Organic catalysts called g._____ do the splicing.

11. Study this figure, which lists the mRNA codons.

First Base	Second Base				Third Base
	U	C	A	G	
U	UUU phenylalanine	UCU serine	UAU tyrosine	UGU cysteine	U
	UUC phenylalanine	UCC serine	UAC tyrosine	UGC cysteine	C
	UUA leucine	UCA serine	UAA stop	UGA stop	A
	UUG leucine	UCG serine	UAG stop	UGG tryptophan	G
C	CUU leucine	CCU proline	CAU histidine	CGU arginine	U
	CUC leucine	CCC proline	CAC histidine	CGC arginine	C
	CUA leucine	CCA proline	CAA glutamine	CGA arginine	A
	CUG leucine	CCG proline	CAG glutamine	CGG arginine	G
A	AUU isoleucine	ACU threonine	AAU asparagine	AGU serine	U
	AUC isoleucine	ACC threonine	AAC asparagine	AGC serine	C
	AUA isoleucine	ACA threonine	AAA lysine	AGA arginine	A
	AUG (start) methionine	ACG threonine	AAG lysine	AGG arginine	G
G	GUU valine	GCU alanine	GAU aspartic acid	GGU glycine	U
	GUC valine	GCC alanine	GAC aspartic acid	GGC glycine	C
	GUA valine	GCA alanine	GAA glutamic acid	GGA glycine	A
	GUG valine	GCG alanine	GAG glutamic acid	GGG glycine	G

a. What does it mean to say that the genetic code is a triplet code?_____

b. What are the mRNA codons for leucine?

_____

12. Using the alphabetized list of terms, label the following diagram to describe translation.

amino acid
anticodon
codon
mRNA
ribosome

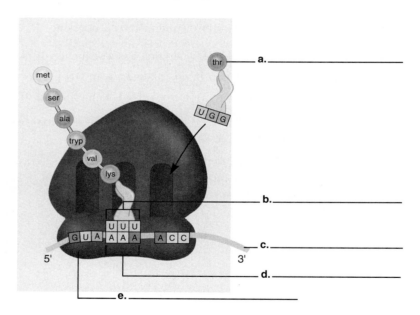

a. _____

b. _____

c. _____

d. _____

e. _____

13. Three types of RNA are present in the diagram for question 12. Ribosomal RNA (rRNA) plus proteins make up the ribosomes. Each ribosome is composed of a(n) ᵃ· _____ subunit and a(n) ᵇ· _____ subunit. Transfer RNA is the second type of RNA in the diagram. At one end a(n) ᶜ· _____ attaches, and at the other end there is a(n) ᵈ· _____, complementary to a codon in mRNA, the third type of RNA.

14. The three steps in translation are ᵃ· _____, when the ribosomal subunits ᵇ· _____; ᶜ· _____, when a polypeptide is ᵈ· _____; and ᵉ· _____, when the last tRNA, the mRNA, and the ribosome ᶠ· _____.

15. Label the following diagram of chain elongation, and fill in the boxes using these descriptions: (1) Two tRNAs can be at a ribosome at one time and the anticodons are paired to the codons; (2) Peptide bond formation attaches the polypeptide chain to the newly arrived amino acid; (3) The ribosome has moved forward, making room for the next incoming tRNA-amino acid complex.

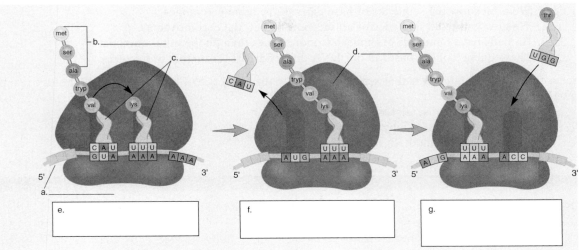

16. As a result of chain elongation, the sequence of ᵃ· _____ in mRNA dictates the order of
    ᵇ· _____ in the polypeptide.

17. For example, if the sequence of bases in mRNA is UUU UUA AUU GUC CCA, the sequence of amino acids in the polypeptide according to the figure in question 11 will be ᵃ· _____ _____. Because several ribosomes, called a(n) ᵇ· _____, can move along one mRNA molecule, ᶜ· _____ (one/many) polypeptides of the same type can be synthesized at a time.

18. Complete the following table:

Levels of Control of Gene Activity	Affects the Activity Of

Use the space provided to answer this question using a complete sentence.

19. What are transcription activators? _____
_____

19.3 GENOMICS (PAGES 378–379)

After you have answered the questions for this section, you should be able to
- Describe the study of genomics.

20. Genomics, the study of ᵃ· _____, has enabled us to sequence all the ᵇ· _____ in all the DNA in all of our ᶜ· _____. The ᵈ· _____ helped to accomplish this feat. There are many similarities between the sequence of our bases and other organisms. We can conclude that we share a large number of
ᵉ· _____ with simpler organisms. The ᶠ· _____ has a goal of cataloging common sequence differences, called ᵍ· _____.

21. Proteomics is the study of the ᵃ· _____, ᵇ· _____, and ᶜ· _____ of cellular proteins.

ᵈ· _____ is the application of computer technologies to the study of the genome.

22. Place a check beside those statements that are true.
    a. _____ We now know the sequence of base pairs in the human genome.
    b. _____ We now know the sequence of all the genes on all the chromosomes.
    c. _____ Humans have hundreds of thousands more genes than do the other animals.
    d. _____ The sequence of the bases in our DNA differs greatly from that of bacteria.

23. Place a check on the statements that are true. Knowing the sequence of bases in our DNA holds great promise for which of these?
    a. _____ Helping to diagnose diseases
    b. _____ Helping to determine why some people get sick
    c. _____ Helping develop methods to treat diseases

19.4 DNA TECHNOLOGY (PAGES 380–384)

After you have answered the questions for this section, you should be able to
- List and describe the means by which genes are cloned.
- Describe the functions of polymerase chain reaction.
- Give examples of biotechnology products produced by bacteria, plants, and animals.

24. In the following diagram, write the numbers of the following descriptions in the appropriate blanks.
    1. Cloning occurs when host cell reproduces.
    2. Host cell takes up recombined plasmid.
    3. DNA ligase seals human gene and plasmid.
    4. Restriction enzyme cleaves DNA.

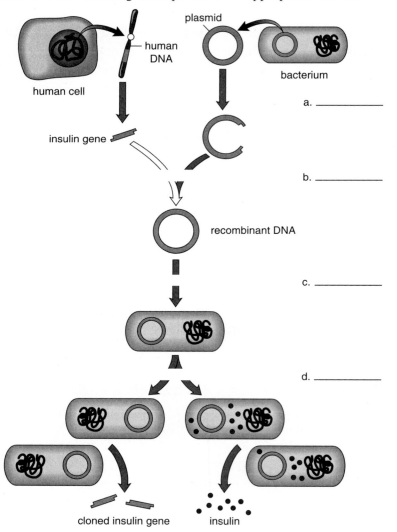

25. What is meant by the expression that restriction enzymes produce "sticky ends"? _____
_____

26. Change the following false statements to true statements:
   a. Plasmids are used as vectors in genetic engineering experiments involving humans. Rewrite: _____
   _____

   b. Recombinant DNA contains two types of bacterial DNA recombined together. Rewrite: _____
   _____

   c. Genetic engineering usually means that an organism receives genes from a member of its own species. Rewrite: ___
   _____

   d. The cloning of a gene occurs when a gene produces many copies of various genes. Rewrite: _____
   _____

27. Explain the polymerase chain reaction by telling what *polymerase* refers to, a._____, and what chain
   reaction means b._____. At the beginning of the reaction, very little DNA may be available, but at
   the end of the reaction c._____ copies of a segment of DNA are available.

28. In DNA fingerprinting, a._____ separates the DNA fragments, and their different lengths are
   compared. If the pattern is identical, the samples are from b._____. Today, DNA fingerprinting often
   utilizes the c._____.

29. Complete the following table on transgenic organisms:

Type of Organism	Engineered for What Purpose?

30. a. The advantage of using bacteria to make a product is that _____.
   b. The advantage of using plants to make a product is that _____.
   c. The advantage of using farm animals to make a product is that _____.

31. Put these statements in the proper sequence (1–5) to describe the production of a transgenic female goat that
   will produce a medicine needed by humans in its milk.
   a. _____ Develop within a host animal.
   b. _____ Remove egg from donor animal.
   c. _____ Isolate a human gene.
   d. _____ Microinject the human gene into the egg of the donor animal.
   e. _____ Transgenic goat is born.

Can you find your way through the maze to a polypeptide by identifying each of the components depicted?

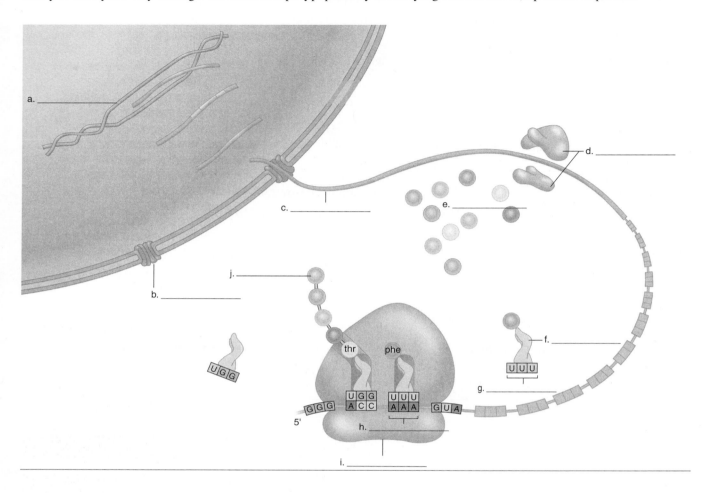

a. _____

b. _____

c. _____

d. _____

e. _____

f. _____

g. _____

h. _____

i. _____

j. _____

If you identified all correctly, you have found your way out.

DEFINITIONS WORDSEARCH

Review key terms by completing this wordsearch using the following alphabetized list of terms:

```
G T R F Y H J D P L O I J
B I O T E C H N O L O G Y
A U Y E H T R A E C D E D
N R T M E W S L A O S N I
T P L P I J Y I R D W E M
I G Y L H F D G C O X Z S
C V B A C X Z A Z N V D A
O T Y T L O P S H Y G F L
D T Y E F R Q E A Z M L P
O T R A N S F E R R N A K
N A P R O T E O M I C S N
```

anticodon
biotechnology
codon
DNA ligase
gene
plasmid
proteomics
template
transfer RNA

a. _____ Unit of heredity passed on to offspring.

b. _____ Enzyme used to join DNA from two sources.

c. _____ RNA molecule that carries the amino acid to the ribosome.

d. _____ Circular fragment of DNA from bacterial cell.

e. _____ Study of the structure, function, and interaction of cellular proteins.

f. _____ Triplet of bases in mRNA.

g. _____ Complementary triplet of bases in tRNA.

h. _____ Use of a natural biological system to produce a commercial product.

i. _____ One strand of DNA that serves as a pattern during duplication.

CHAPTER TEST

OBJECTIVE TEST

Do not refer to the text when taking this test.

____ 1. Which is NOT true of DNA?
  a. polymer of nucleotides
  b. occurs in nucleus
  c. has a varied sequence of just three bases
  d. contains deoxyribose

____ 2. DNA is the hereditary material and it
  a. stores genetic information.
  b. replicates.
  c. controls cells' activities.
  d. All of these are correct.

____ 3. Which of the following pairs is NOT a valid comparison between DNA and RNA?

DNA	RNA
a. double helix	single stranded
b. replicates	duplicates
c. deoxyribose	ribose
d. thymine	uracil

____ 4. A person with a metabolic defect, such as the inability to make a particular protein, inherited a
  a. faulty gene.
  b. faulty code.
  c. mutation.
  d. All of these are correct.

_____ 5. If a DNA molecule has this sequence of base pairs, what will be the sequence following replication?

```
┌── A ········ T ──┐
├── T ········ A ──┤
├── C ········ G ──┤
└── G ········ C ──┘
```

    a. exactly the same as this
    b. exactly opposite of this
    c. wherever a *T* appears after replication there would be a *U*
    d. like this, only each DNA molecule would be single stranded

_____ 6. If mutation occurred, then
    a. the code would change.
    b. some particular codon or codons would change.
    c. some particular anticodon or anticodons would change.
    d. All of these are correct.
    e. Both *a* and *b* are correct.

_____ 7. Transcription
    a. involves DNA.
    b. involves tRNA.
    c. occurs at the ribosome.
    d. All of these are correct.

_____ 8. A mutation has occurred when the sequence of
    a. bases in DNA changes.
    b. bases in mRNA changes.
    c. amino acids in protein changes.
    d. All of these are correct.

_____ 9. Which pair is mismatched?
    a. DNA—code
    b. mRNA—codon
    c. tRNA—anticodon
    d. rRNA—recodon

_____ 10. Which is the correct order of the numbered phrases to describe gene expression?
  (1) mRNA is produced in the nucleus.
  (2) Ribosomes move along mRNA.
  (3) DNA has a code.
  (4) Polypeptide results.
  (5) tRNA brings amino acids to ribosomes.
  (6) mRNA moves to ribosomes.
    a. 1,2,3,4,5,6
    b. 2,4,6,1,3,5
    c. 3,1,6,2,5,4
    d. 3,5,1,6,2,4

_____ 11. Protein synthesis takes place in the
    a. nucleus, where DNA codes for amino acids.
    b. cytoplasm, where the ribosomes code for amino acids.
    c. nucleus, where the tRNAs code for amino acids.
    d. cytoplasm, where mRNA codons pair with tRNA anticodons.

_____ 12. How does the anticodon differ from the codon?
    a. The anticodon contains thymine, but the codon contains uracil.
    b. The anticodon attaches to ribosomes, but the codon attaches to amino acids.
    c. The anticodon is a sequence of three bases complementary to the bases of a codon.
    d. The anticodon stands for a particular amino acid, but the codon codes for nucleotide bases.
    e. All of these are correct.

_____ 13. What two types of molecules are involved when the codon pairs with its anticodon?
    a. mRNA and DNA
    b. mRNA and tRNA
    c. DNA and tRNA
    d. rRNA and DNA

_____ 14. Which of these is happening when translation takes place?
    a. mRNA is still in the nucleus.
    b. tRNAs are bringing amino acids to the ribosomes.
    c. rRNAs expose their anticodons.
    d. DNA is being replicated.
    e. All of these are correct.

_____ 15. Which two events are in proper sequence?
    a. translation—replication
    b. translation—transpiration
    c. transcription—translation
    d. translation—transcription

_____ 16. Which of these is a true statement concerning translation?
    a. Each polypeptide is synthesized one amino acid at a time.
    b. The amino acids are joined by RNA polymerase at the same time.
    c. Each ribosome is responsible for adding a single amino acid to each polypeptide.
    d. Because the code is degenerate, the same type of polypeptide often contains a different sequence of amino acids.
    e. All of these are correct.

_____ 17. Replication is considered
    a. semiconservative.
    b. conservative.
    c. termination.
    d. initiation.

_____ 18. Plasmids
  a. are found in bacteria.
  b. are rings of DNA.
  c. may be used to make recombinant DNA.
  d. All of these are correct.
_____ 19. A biotechnology product could be a(n)
  a. enzyme.
  b. vaccine.
  c. hormone.
  d. All of these are correct.
_____ 20. To make a transgenic animal, you have to have
  a. an egg donor.
  b. special food to feed it.
  c. a way to protect human beings from coming in contact with transgenic animals.
  d. a host animal.
  e. both *a* and *d*.
_____ 21. The polymerase chain reaction
  a. produces many copies of different segments of DNA.
  b. produces many copies of the same segment of DNA.
  c. requires denaturing double-stranded DNA into single strands by heating.
  d. Both *b* and *c* are correct.

_____ 22. Which of these is NOT a step in preparing recombinant DNA?
  a. Remove plasmid from bacterial cell.
  b. Use restriction enzyme to acquire foreign gene and cut open vector.
  c. Use ligase to seal foreign gene into vector.
  d. Use a virus to carry recombinant DNA into a plasmid.
_____ 23. Which of the following chemicals now on the market have been produced by biotechnology?
  a. hepatitis B vaccine
  b. t-PA
  c. insulin
  d. All of these are correct.
_____ 24. Transgenic plants can be made
  a. resistant to pesticides.
  b. by introducing genes into full-grown plants.
  c. resistant to herbicides.
  d. Both *a* and *c* are correct.
_____ 25. The human genome
  a. will probably be completely sequenced in 50 years.
  b. has no usefulness to humans.
  c. contains only about 30,000 genes.
  d. is known only to government employees because it is top secret.

THOUGHT QUESTIONS

Use the space provided to answer these questions in complete sentences.
26. It is said that DNA stores information. What kind of information does it store?

27. Why would you expect a transgenic animal to pass the newly acquired gene of interest on to its offspring?

Test Results: _____ number correct ÷ 27 = _____ × 100 = _____%

STUDY QUESTIONS

1. a. T **b.** T **c.** F . . . composed of the sugar deoxyribose **d.** T **2. a.** phosphate, sugar **b.** nitrogen-containing bases **c.** A is paired with T and G is paired with C and vice versa. **d.** hydrogen **e.** twist it **3. a.** hydrogen bond **b.** they become unzipped **c.** complementary base pairing **d.** DNA polymerase **e.** old, new, semiconservative **f.** the same **4. a.** nucleus **b.** cytoplasm **c.** RNA **5. a.** T **b.** F **c.** F **d.** F **e.** T **6. a.** is found in ribosomes, where proteins are synthesized **b.** takes a message from DNA in the nucleus to the ribosomes in the cytoplasm **c.** transfers amino acids to the ribosomes **7. a.** amino acid **b.** amino acids **c.** no **d.** no **e.** sequence of amino acids **f.** as enzymes **8. a.** transcription **b.** translation **9. a.** bases **b.** amino acid **c.** codons **10. a.** complementary **b.** U **c.** C **d.** GCU UCG AGA **e.** The code is a triplet code, and each codon contains three bases. **f.** introns **g.** ribozymes **11. a.** Every three bases stand for an amino acid. **b.** UUA, UUG, CUU, CUC, CUA, CUG **12. a.** amino acid **b.** anticodon **c.** mRNA **d.** codon **e.** ribosome **13. a.** large **b.** small **c.** amino acid **d.** anticodon **14. a.** initiation **b.** associate **c.** elongation **d.** lengthened **e.** termination **f.** dissociate **15. a.** mRNA **b.** polypeptide **c.** tRNA **d.** ribosome **e.** (1) Two tRNAs . . . **f.** (2) Peptide bond . . . **g.** (3) The ribosome . . . **16. a.** bases **b.** amino acids **17. a.** phenylalanine, leucine, isoleucine, valine, proline **b.** polyribosomes **c.** many

18.

Levels of Control of Gene Activity	Affects the Activity of
transcriptional	DNA
posttranscriptional	mRNA during formation and processing
translational	mRNA life span during protein synthesis
posttranslational	protein

19. Factors that must bind to DNA before transcription can begin. **20. a.** genomes **b.** base pairs **c.** chromosomes **d.** Human Genome Project **e.** genes **f.** HapMap project **g.** haplotypes **21. a.** structure **b.** function **c.** interaction **d.** Bioinformatics **22.** a **23.** a, b, c **24. a.** 4 **b.** 3 **c.** 2 **d.** 1 **25.** Cleavage results in unpaired bases. **26. a.** . . . involving bacteria **b.** . . . contains DNA from two different sources **c.** . . . from a member of a different species **d.** . . . occurs when many copies of the same gene are produced. **27. a.** DNA polymerase, the enzyme involved in DNA replication **b.** the reaction occurs over and over again **c.** many **28. a.** gel electrophoresis **b.** from the same individual **c.** polymerase chain reaction

29.

Type of Organism	Engineered for What Purpose?
bacteria	to make products to protect plants, to clean up the environment, to produce chemicals, and to mine metals
plants	to resist insects, pesticides and herbicides, and to make products
animals	to have improved qualities and to make products

30. a. they will take up plasmids **b.** they will grow from single cells (protoplasts) **c.** the product is easily obtainable in milk **31.** c, b, d, a, e

GENE EXPRESSION MAZE

a. DNA **b.** nuclear pore **c.** mRNA **d.** ribosomal subunits **e.** amino acids **f.** tRNA **g.** anticodon **h.** codon **i.** ribosome **j.** polypeptide chain

DEFINITIONS WORDSEARCH

```
                    D
    B I O T E C H N O L O G Y
    A     E     A   C   E D
    N     M     L   O   N I
    T     P     I   D   E M
    I     L     G   O     S
    C     A     A   N     A
    O     T     S         L
    D     E     E         P
    O T R A N S F E R R N A
    N   P R O T E O M I C S
```

a. gene **b.** DNA ligase **c.** transfer RNA **d.** plasmid **e.** proteomics **f.** codon **g.** anticodon **h.** biotechnology **i.** template

CHAPTER TEST

1. c **2.** d **3.** b **4.** d **5.** a **6.** e **7.** a **8.** d **9.** d **10.** c **11.** d **12.** c **13.** b **14.** b **15.** c **16.** a **17.** a **18.** d **19.** d **20.** e **21.** d **22.** d **23.** d **24.** d **25.** c **26.** The sequence of bases in DNA dictates the proper sequence of amino acids in a protein. This is the information stored in DNA. **27.** The gene is now a part of the animal's chromosomes and is passed on with the chromosomes.

20

GENETIC COUNSELING

It is possible to look at the chromosomes of a cell during metaphase of mitosis (karyotype). This is useful for determining whether or not a chromosomal abnormality exists. Amniocentesis and chorionic villi sampling can provide fetal cells for the karyotype.

Note the steps for obtaining cells for a karyotype and how the chromosomes are arranged (pp. 390–391). Make a chart of the various genetic diseases and what alteration has occurred from the normal karyotype for each (pp. 392–397).

Make a key to use for interpreting the various symbols used in a pedigree (p. 398), and write down the characteristics of an autosomal dominant, an autosomal recessive, and an X-linked disorder pedigree (pp. 398–399), so that you can distinguish between these disorders.

Make a chart of each of the single gene mutation disorders, how they are inherited (autosomal versus X-linked, dominant versus recessive), and the characteristic symptoms for the disease (pp. 400–402). The key to keeping all of the genetic diseases organized in your thoughts is to review your chart frequently.

Know how to test for a genetic using disorder either protein or DNA (p. 403), from a fetus, an embryo, or an egg (pp. 404–405).

Be able to discuss the benefits and disadvantages of genetic profiling and gene therapy (pp. 406–409). You will want to be informed of these, since they will be involved in various pieces of federal legislation.

TIP: Be sure to visit the Online Learning Center that accompanies *Human Biology* 9/e. It has practice quizzes, interactive activities, labeling exercises, art quizzes, animators, flash cards, and much more. http://www.mhhe.com/maderhuman9

Study the text section by section. Answer the study questions so that you can fulfill the learning objectives for each section.

20.1 COUNSELING FOR CHROMOSOMAL DISORDERS (PAGES 390–397)

After you have answered the questions for this section, you should be able to
- Describe a normal karyotype of a human being.
- Explain the purpose of amniocentesis and chorionic villi sampling.
- Draw diagrams showing nondisjunction during meiosis I and meiosis II.
- Describe syndromes resulting from inheritance of an abnormal chromosome number.
- Describe chromosomal mutations, including deletions, duplications, translocations, and inversions.
- Describe the various syndromes associated with chromosomal mutations.

1. Place *A* for amniocentesis and *C* for chorionic villi sampling on the appropriate lines.
   a. _____ 14 to 17 weeks of pregnancy.
   b. _____ 5 weeks of pregnancy.
   c. _____ Cells are obtained by suction.
   d. _____ Cells are obtained by needle.
   e. _____ Cells are from a cavity around the embryo/fetus.
   f. _____ Cells are from the extraembryonic membrane itself.

2. Match the descriptions with these terms:

   X and Y chromosomes   XX   homologous chromosomes   autosomes   XY

   a. _____ sex chromosomes
   b. _____ all chromosomes but the sex chromosomes
   c. _____ pairs of chromosomes
   d. _____ female
   e. _____ male

3. Label the following photograph, using these alphabetized terms:

   autosomes   homologous pair   karyotype of a male   sex chromosomes

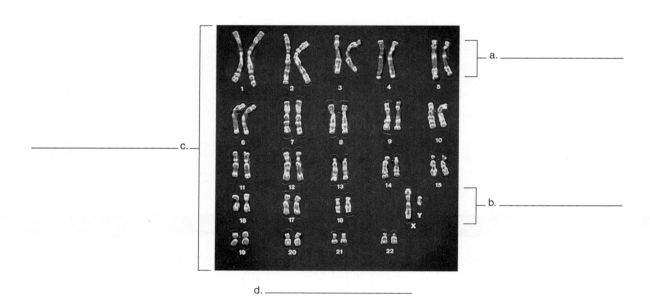

a. _____

c. _____

b. _____

d. _____

4. a. Down syndrome arises when the egg has two copies of chromosome 21. Complete the following diagram by adding chromosomes to illustrate the occurrence of nondisjunction during meiosis II (left-hand side) and nondisjunction during meiosis I (right-hand side) to produce an individual with Down syndrome.

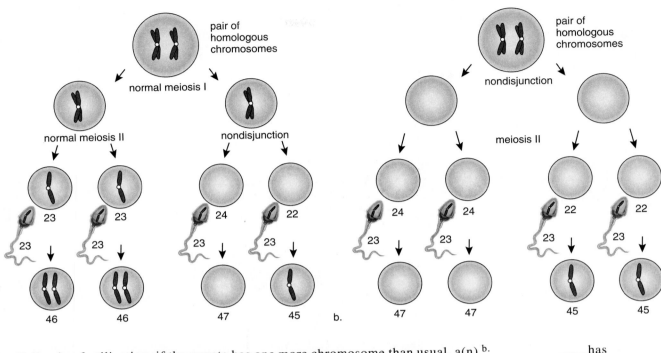

Following fertilization, if the zygote has one more chromosome than usual, a(n) b. _____ has occurred. On the other hand, if the zygote has one less chromosome than usual, a(n) c. _____ has occurred.

5. Since males are XY and females are XX, the gender of the individual is normally determined by the
   a. _____ parent, depending on whether the offspring receives a(n) b. _____ or a(n) c. _____ chromosome.

6. Match the following conditions to each of the descriptions below. Terms can be used more than once.
   1. Turner syndrome
   2. Klinefelter syndrome
   3. poly-X individual
   4. Jacobs syndrome
   a. _____ female with no apparent physical abnormalities
   b. _____ male with some breast development, large hands
   c. _____ XYY male
   d. _____ XXY male
   e. _____ XXX female
   f. _____ XO female
   g. _____ female with no Barr body

7. Identify the types of chromosomal mutations shown in the following illustration.

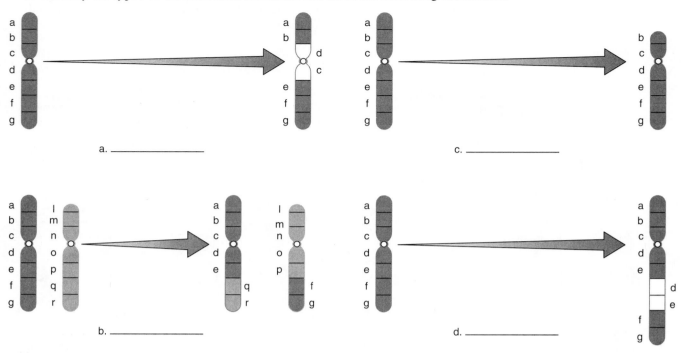

a. _____

c. _____

b. _____

d. _____

Cri du chat syndrome is due to which of these mutations? e. _____

20.2 COUNSELING FOR GENETIC DISORDERS: THE PRESENT (PAGES 398–405)

After you have answered the questions for this section, you should be able to
- Make use of a pedigree for both dominant and recessive genetic disorders.
- Describe the autosomal dominant and autosomal recessive disorders discussed in the text.
- Be able to solve genetics problems concerning autosomal dominant and autosomal recessive disorders.

In questions 8–9, fill in the blanks.

8. Following is a portion of a pedigree for a dominant disorder. Shaded individuals are affected.

Which of these individuals could be *AA?* a. _____ Explain: _____

_____

Which of these individuals is known to be *Aa?* b. _____ Explain: _____

_____

Which of these is *aa?* c. _____ Explain: _____

_____

Are heterozygotes affected or unaffected? d. _____

9. Following is a portion of a pedigree for a recessive disorder.

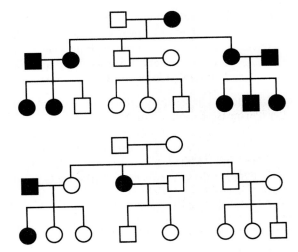

Which of these individuals could be *AA?* a. _____ Explain: _____
_____

Which of these individuals is *Aa?* b. _____ Explain: _____
_____

Which of these individuals is *aa?* c. _____ Explain: _____
_____

Are homozygous dominant individuals and heterozygous individuals affected or unaffected? d. _____
_____

For questions 10–11, study the pedigrees.

10. Decide whether the condition is dominant or recessive. Indicate the genotype of each person in the pedigree using the letters *A* and *a*.

11. Decide whether the condition is dominant or recessive. Indicate the genotype of each person in the pedigree using the letters *A* and *a*.

12. Match the following disorders to each of the statements.
   1. Marfan syndrome
   2. Huntington disease
   3. Tay-Sachs disease
   4. cystic fibrosis
   5. phenylketonuria
   6. sickle-cell disease
   7. sickle-cell trait
   a. _____ a defect in an elastic connective tissue protein, called fibrillin
   b. _____ accumulation of phenylalanine in urine
   c. _____ blood cells are normal unless they experience dehydration or mild O_2 deprivation
   d. _____ brain cell degeneration
   e. _____ red blood cells are sickle-shaped and clog arteries
   f. _____ buildup of mucus in respiratory system
   g. _____ buildup of glycosphingolipid in lysosomes

13. A person heterozygous for Huntington disease reproduces with a person who is perfectly normal. What are the chances of an offspring developing Huntington disease when older? _____

14. Mary is 50 years old and has Huntington disease. Her father was killed accidentally at a young age; her mother is 80 years old and is normal. What is the most likely genotype of all persons involved? Use the alleles *A* and *a*.

    a. Mary's mother _____

    b. Mary's father _____

    c. Mary _____

15. Both parents appear to be normal, but their child has cystic fibrosis. What are the genotypes of all persons involved? Use the alleles *A* and *a*.

    a. parents _____   b. child _____

20.3 COUNSELING FOR GENETIC DISORDERS: THE FUTURE (PAGES 406–408)

After you have answered the questions in this section, you should be able to
- Discuss what we now know about the human genome and how this knowledge may benefit humans.
- Describe two possible methods for gene therapy in humans.

16. Place a check beside those statements that are true.
    _____ a. We now know the sequence of base pairs in the human genome.
    _____ b. We know how each person's genome differs from the norm.
    _____ c. Researchers are working on linking abnormalities to base sequence changes.
    _____ d. DNA chips hold all the genes in our genome.

17. Place a check beside the statements that are true. Knowing the sequence of bases in our DNA holds great promise for which of these?
    _____ a. Helping locate mutations in a person's genome
    _____ b. Helping determine why some people get sick more than others
    _____ c. Helping develop drugs that are specific for the various causes of the same type of disorder

18. Change these false statements to true statements.

    a. Ex vivo methods of gene therapy require that the therapeutic gene be placed in the body either directly or by using a viral vector. Rewrite: _____

    _____

    b. A common ex vivo method is to microinject normal genes into bone marrow stem cells removed from the patient. Then the stem cells are returned to the patient. Rewrite: _____

    _____

    c. Gene therapy is currently restricted to curing genetic diseases and is not used to treat illnesses such as cystic fibrosis or cardiovascular diseases. Rewrite: _____

    _____

Review key terms by completing this matching exercise, selecting from the following alphabetized list of terms:

amniocentesis
Barr body
chromosomal mutation
karyotype
nondisjunction
trisomy

a. _____ This appears in a cell's nucleus as a dark staining structure due to the inactivation of an X-chromosome.

b. _____ Failure of homologues or sister chromatids to separate during the formation of gametes.

c. _____ Variation in regard to the normal number of chromosomes inherited or in regard to the structure there of.

d. _____ Arrangement of all the chromosomes within a cell by pairs in a fixed order.

e. _____ Procedure for removing amniotic fluid surrounding the developing fetus for the testing of the fluid or cells within the fluid.

f. _____ This occurs when a sperm with one of a particular chromosome fertilizes an egg containing two of that chromosome.

CHAPTER TEST

OBJECTIVE QUESTIONS

Do not refer to the text when taking this test.

_____ 1. Which phrase best describes the human karyotype?
   a. 46 pairs of autosomes
   b. one pair of sex chromosomes and 23 pairs of autosomes
   c. X and Y chromosomes and 22 pairs of autosomes
   d. one pair of sex chromosomes and 22 pairs of autosomes

_____ 2. The gene arrangement on a chromosome changes from *ABCDEFG* to *ABCDEDEFG*. This is an example of
   a. deletion.
   b. duplication.
   c. inversion.
   d. linkage.

_____ 3. Which chromosomal mutation does NOT require the presence of another chromosome?
   a. translocation
   b. duplication
   c. inversion
   d. Both *b* and *c* are correct.

_____ 4. Which type of chromosomal mutation occurs when two simultaneous breaks in a chromosome lead to the loss of a segment?
   a. inversion
   b. translocation
   c. deletion
   d. duplication

_____ 5. Which of these is an autosomal abnormality?
   a. Turner syndrome
   b. Down syndrome
   c. Klinefelter syndrome
   d. poly-X syndrome

_____ 6. XYY (Jacobs syndrome) males occur due to nondisjunction during
   a. oogenesis.
   b. spermatogenesis.
   c. fertilization.
   d. mitosis.

_____ 7. Which condition is more likely to occur when the mother is over age 40?
   a. Turner syndrome
   b. poly-X syndrome
   c. Down syndrome
   d. Klinefelter syndrome

Questions 8–9 pertain to this pedigree chart.

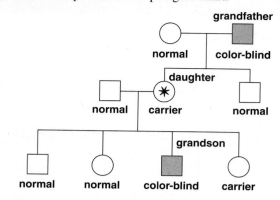

_____ 8. The allele for this disorder is
  a. dominant.
  b. recessive.
  c. X-linked and recessive.
  d. None of these is correct.

_____ 9. The genotype of the starred individual is
  a. *Aa.*
  b. *aa.*
  c. $X^A X^a.$
  d. $X^A Y^a.$

_____ 10. A woman who is a carrier for color blindness reproduces with a man who has normal color vision. What is the chance they will have a color-blind daughter?
  a. 50%
  b. 25%
  c. 100%
  d. no chance

_____ 11. A color-blind woman reproduces with a man who has normal color vision. Their sons will
  a. be like the father because the trait is X-linked.
  b. be like the mother because the trait is X-linked.
  c. all have normal color vision.

_____ 12. A girl is color-blind.
  a. She received an allele for color blindness from her mother.
  b. She received an allele for color blindness from her father.
  c. All her sons will be color-blind.
  d. Her father is color-blind.
  e. All of these are correct.

_____ 13. Which chromosome has genes to determine male genital development?
  a. chromosome 5
  b. chromosome 10
  c. chromosome 21
  d. X chromosome
  e. Y chromosome

_____ 14. The _____ genotype indicates carrier female for color blindness.
  a. $X^B X^B$
  b. $X^B X^b$

  c. $X^b X^b$
  d. $X^B Y$
  e. $X^b Y$

_____ 15. Genomics
  a. is the study of cellular protein structure.
  b. sequences mutant genes.
  c. is the study of the human genome.
  d. is a mechanism used in DNA fingerprinting.

_____ 16. Which of the following statements is NOT correct?
  a. Females have an XX genotype.
  b. Males have an XY genotype.
  c. An egg always bears an X chromosome.
  d. Each sperm cell has an X and Y chromosome.
  e. The sex of the newborn child is determined by the father.

_____ 17. Which of the following genotypes is not paired with the correct number of Barr bodies?
  a. XY—none
  b. XO—one
  c. XX—one
  d. XXY—one
  e. XXX—two

_____ 18. Which of the following is NOT a characteristic of an X-linked recessive disorder?
  a. More males than females are affected.
  b. An affected son can have parents who have the normal phenotype.
  c. For a female to have the characteristic, her father must also have it.
  d. The characteristic often skips a generation from the grandmother to the granddaughter.
  e. If a woman has the characteristic, all of her sons will have it.

_____ 19. Which of the following is NOT considered an X-linked recessive disorder?
  a. Tay-Sachs disease
  b. color blindness
  c. hemophilia
  d. muscular dystrophy (some forms)

_____ 20. What does the Human Genome Project have to do with gene therapy?
  a. Knowing the location of all the genes on the chromosomes will make it easier to isolate particular genes to cure humans.
  b. Both the Human Genome Project and gene therapy are unethical and therefore, will never be completed.
  c. Both the Human Genome Project and gene therapy require the use of restriction enzymes and gel electrophoresis.
  d. Both the Human Genome Project and gene therapy require prior use of the polymerase chain reaction.
  e. All but *b* are correct.

21. If a cell is altered while outside the human body for gene therapy, it is considered _____ therapy.
   a. ex vivo
   b. in vivo
   c. in vitro
   d. extraneous
   e. intravenous

_____22. The X-linked disease prevalent among royal families of Europe at the turn of the twentieth century was
   a. muscular dystrophy.
   b. color blindness.
   c. hemophilia.
   d. Williams syndrome.
   e. cri du chat syndrome.

THOUGHT QUESTIONS

Answer in complete sentences.

23. Why is it evident that a gene for maleness exists on the Y chromosome?

24. Why are color-blind women rare?

Test Results: _____ number correct ÷ 24 = _____ × 100 = _____ %

ANSWER KEY

STUDY QUESTIONS

1. a. A **b.** C **c.** C **d.** A **e.** A **f.** C   **2. a.** X and Y chromosomes **b.** autosomes **c.** homologous chromosomes **d.** XX **e.** XY   **3. a.** homologous pair **b.** sex chromosomes **c.** autosomes **d.** karyotype of a male   **4. a.** See Fig. 20.2, p.392, in text. **b.** trisomy **c.** monosomy   **5. a.** male **b.** X **c.** Y   **6. a.** 3 **b.** 2 **c.** 4 **d.** 2 **e.** 3 **f.** 1 **g.** 1   **7. a.** inversion **b.** translocation **c.** deletion **d.** duplication **e.** deletion   **8. a.** None. Both parents are affected, but they have an unaffected child. **b.** Both parents are *Aa*. Otherwise, they could not have an unaffected child. **c.** individual 3 because this individual is unaffected **d.** affected   **9. a.** None; individuals 1 and 2 are not shaded, but they are heterozygous because they have an affected child **b.** 1 and 2; they have to be heterozygous in order to have an affected child **c.** Individual 3; only the recessive can have parents that are unaffected **d.** unaffected   **10.** dominant **top:** *aa, Aa* **middle:** *Aa, Aa; aa; aa; Aa A?* **bottom:** *A? A? aa; aa, aa, aa; A?; A?, A?*   **11.** recessive **top:** *Aa, Aa* **middle:** *aa, Aa; aa, A?; A?, A?* **bottom:** *aa, Aa, Aa; Aa, Aa; A?, A?, A?*   **12. a.** 1 **b.** 5

c. 7 **d.** 2 **e.** 6 **f.** 4 **g.** 3   **13.** 50%   **14. a.** *aa* **b.** *Aa* **c.** *Aa*   **15. a.** *Aa* **b.** *aa*   **16.** a   **17.** a, b, c   **18. a.** In vivo **b.** . . . use a viral vector to carry normal genes . . . **c.** . . . is not restricted to curing genetic disease and is used to treat illnesses . . .

DEFINITIONS WORDMATCH

a. Barr body   **b.** nondisjunction   **c.** chromosomal mutation   **d.** karyotype   **e.** amniocentesis   **f.** trisomy

CHAPTER TEST

1. d   **2.** b   **3.** d   **4.** c   **5.** b   **6.** b   **7.** c   **8.** c   **9.** c   **10.** d   **11.** b   **12.** e   **13.** e   **14.** b   **15.** a   **16.** d   **17.** b   **18.** d   **19.** a   **20.** a   **21.** a   **22.** c   **23.** Any individual receiving a Y chromosome is male.   **24.** A color-blind woman has to receive an allele for color blindness from both parents. If she receives only one allele for color blindness and one normal allele, she will not be color-blind.

PART 7 HUMAN DISEASE

21

DEFENSES AGAINST DISEASE

STUDY TIPS

How the body defends itself from disease is one of the most fascinating areas of human biology. In this chapter, you will learn how the immune system, which includes the lymphatic organs and white blood cells, protects us from disease.

Make a list of the lymphatic organs and their functions (pp. 414–415). Understand the division of lymphatic organs into those that are primary and those that are secondary.

Also make a list of the white blood cells (pp. 416–423) and their functions in immunity. Understand the difference between nonspecific and specific defenses. Understand, as well, that the MHC (p. 422) are self antigens that all the cells of the immune system use to recognize each other and the other cells of your body.

Also, molecules of interest include those of the complement system (p. 418), the cytokines (p. 425), and monoclonal antibodies (pp. 425–426). By doing some careful "bookkeeping" of terms in this chapter, you will not confuse them with each other.

Immunity side effects include allergies and various autoimmune diseases, which plague many persons at the present time (pp. 426–428).

TIP: Be sure to visit the Online Learning Center that accompanies *Human Biology* 9/e. It has practice quizzes, interactive activities, labeling exercises, art quizzes, animations, flash cards, and much more.
http://www.mhhe.com/maderhuman9

STUDY QUESTIONS

Study the text section by section. Answer the study questions so that you can fulfill the learning objectives for each section.

21.1 ORGANS, TISSUES, AND CELLS OF THE IMMUNE SYSTEM (PAGES 414–415)

After you have answered the questions for this section, you should be able to
- Describe the location and function of the primary and secondary lymphatic organs.
- Discuss the functions of the spleen and of the white and red pulp it contains.
- Describe how the thymus gland and red bone marrow participate in immunity.
- Explain the structure and purpose of lymph nodes.

1. Place the appropriate letter next to each organ.

   P—primary lymphatic organ   *S*—secondary lymphatic organ

   a. _____ spleen
   b. _____ thymus gland
   c. _____ red bone marrow
   d. _____ lymph nodes
2. Place the appropriate letter(s) next to each statement.

   T— thymus   *S*—spleen   *RBM*—red bone marrow

   a. _____ contains red pulp and white pulp
   b. _____ contains stem cells
   c. _____ is located posterior to the sternum in the upper thoracic cavity
   d. _____ is located in the upper left abdominal cavity

3. Place the appropriate letter(s) next to each description.

   L— lymph nodes   *RBM*—red bone marrow   *S*—spleen   *T*—Thymus

   a. _____ maturity of T lymphocytes
   b. _____ production of all blood cells
   c. _____ filters the blood
   d. _____ filters the lymph
   e. _____ produces hormones believed to stimulate the immune system

4. Patches of lymphatic tissue occur in specific locations.

   a. Where are the tonsils located? _____

   b. Where are Peyer's patches located? _____

   c. Where is the appendix located? _____

21.2 NONSPECIFIC AND SPECIFIC DEFENSES (PAGES 417–423)

After you have answered the questions for this section, you should be able to
- Describe barriers to entry, the inflammatory reaction, natural killer cells, and protective proteins as agents of nonspecific immunity.
- Compare the origin, maturation, and function of B cells and T cells.
- Explain how and where B cells undergo clonal selection and expansion.
- Describe the general structure of an antibody, and state a function for the variable and constant regions.
- Tell how a T cell recognizes an antigen.
- List the different types of T cells and give the action of each.

5. Match the descriptions to these defense mechanisms:

   1. barrier to entry   2. inflammatory response   3. complement system   4. natural killer cells

   a. _____ accompanied by swelling and redness
   b. _____ cilia action in the respiratory tract
   c. _____ produces holes in bacterial cell walls
   d. _____ stomach secretions
   e. _____ histamine increases capillary permeability
   f. _____ the production of pus
   g. _____ vagina is inhabited by nonpathogenic bacteria
   h. _____ stimulates inflammatory response, binds to some bacteria and punches holes in others
   i. _____ neutrophils and macrophages carry out phagocytosis
   j. _____ secretions of the oil, or sebaceous, glands
   k. _____ kills cells infected with a virus and tumor cells

6. Fill in this table with *yes* or *no:*

	Cytotoxic T Cell	B Cell
Ultimately derived from stem cells in bone marrow	a.	b.
Pass through thymus	c.	d.
Carry antigen receptors on membrane	e.	f.
Cell-mediated immunity	g.	h.
Antibody-mediated immunity	i.	j.

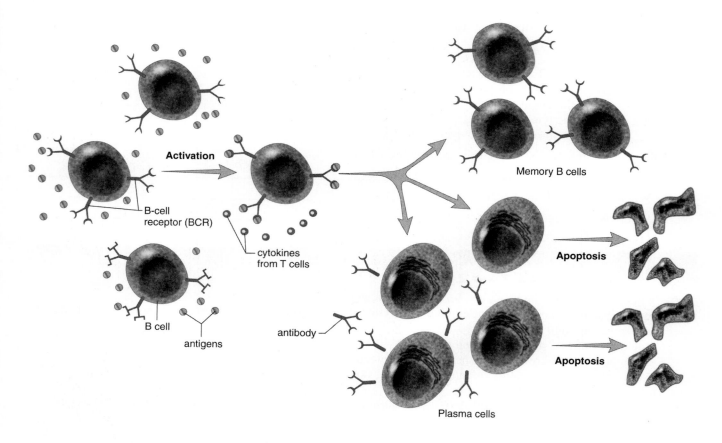

B-cell
receptor (BCR)

Activation

cytokines
from T cells

B cell

antigens

antibody

Memory B cells

Apoptosis

Apoptosis

Plasma cells

Study the diagram above and answer questions 7–10.

7. Which of these makes a B cell undergo clonal expansion?
   a. when an antigen binds to its antigen receptor
   b. when fever is present

8. Explain the expression clonal selection theory. _____
   _____

9. Which of these are cells that result from the clonal expansion of B cells?
   a. plasma cells
   b. memory cells
   c. Both types of cells.

10. What happens to these cells once the infection is under control?
    a. memory B cells _____
    b. plasma cells _____

11. Using these terms, label this diagram of an antibody molecule:
    antigen-binding site
    constant region
    heavy chain
    light chain
    variable region

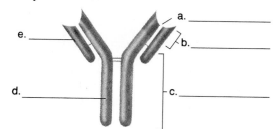

e. _____

a. _____

b. _____

d. _____

c. _____

    f. What is the function of antibodies? _____

Study the following diagram and answer questions 12–14.

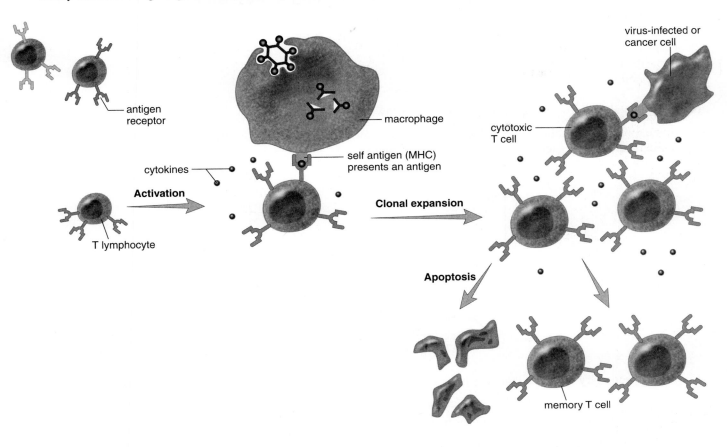

antigen receptor

macrophage

cytokines

Activation

self antigen (MHC) presents an antigen

T lymphocyte

Clonal expansion

virus-infected or cancer cell

cytotoxic T cell

Apoptosis

memory T cell

12. Which of these makes a particular T cell undergo clonal expansion?
    a. when the antigen is presented to the T cell by an APC
    b. when a T cell encounters an antigen

13. a. What is the significance of MHC proteins? _____
    _____

14. a. What happens after a helper T cell recognizes an antigen? _____
    _____

    b. What happens after a cytotoxic T cell recognizes an antigen? _____
    _____

    c. What happens to T cells (except for memory T cells) after the infection is past? _____
    _____

21.3 ACQUIRED IMMUNITY (PAGES 424–426)

After you have answered the questions for this section, you should be able to
- Describe the primary and secondary responses to an immunization.
- Differentiate between active and passive immunity.
- Describe the role of cytokines in immunity.
- Define monoclonal antibody.

15. Using the alphabetized list of terms, label the following diagram.

first exposure to vaccine   plasma antibody concentration   primary response
second exposure to vaccine   secondary response

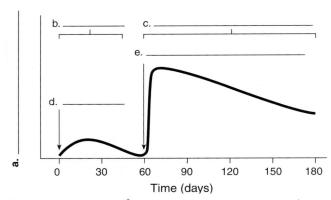

A good secondary response can be related to f. _____, dependent on the number of

g. _____ and h. _____ cells capable of responding to a particular antigen.

16. When an individual receives antibodies from another, as when a baby breast-feeds, it is called _____.

17. List five diseases for which children should be immunized. _____

_____

18. Why is it better to immunize children in advance than to wait until the disease has been contracted?

_____

19. Name two types of cytokines, and tell what role they play in immunotherapy.

a. _____

b. _____

20. What are monoclonal antibodies? _____

_____

21.4 IMMUNITY SIDE EFFECTS (PAGES 426–428)

After you have answered the questions for this section, you should be able to
- List three types of immunity side effects and relate them to the function of the immune system.

21. Indicate whether these statements are true (T) or false (F).
   a. _____ An allergy to pollen would be classified as an immediate allergic response.
   b. _____ The most frequently prescribed immunosuppressive drugs act by inhibiting the response of T cells to cytokines.
   c. _____ Multiple sclerosis and many other conditions are now known to be autoimmune diseases.

22. Describe the immune response for each of these:

   a. allergy _____

   b. tissue rejection _____

   c. autoimmune disease _____

DEFINITIONS WORDSEARCH

Review key terms by completing this wordsearch using the following alphabetized list of terms:

```
L E N I M A T S I H E R T M
I G S D C O M P L E M E N T
E A N T I B O D Y D L P A O
N H A N T I G E N E S H U D
I P H L A A R N O M P U T V
K O C E P D H E R A L O O B
O R M C O Y K D E H E L I Y
H C E E P K U F F P E E M M
P A P P T I O R R S N V M B
M M M U O N G H E E M Y U D
Y Y D O S I E T T S H I N C
L E A S I N T Y N P P L E C
M U D S S I D E I Y H N T D
```

antibody
antigen
apoptosis
autoimmune
complement
cytokine
histamine
interferon
macrophage
spleen

a. _____ Foreign protein.

b. _____ System of plasma proteins that are a nonspecific defense.

c. _____ Disease caused by the immune system attacking the person's own body.

d. _____ Programmed cell death.

e. _____ Molecule released by T cells that enhances the abilities of other immune system cells.

f. _____ Large phagocytic cell.

g. _____ Protein released by cells infected with viruses.

h. _____ Protein produced by B cells in response to foreign antigens.

i. _____ Substance produced by damaged tissue cells or mast cells in an allergic reaction.

j. _____ Lymphatic organ that filters blood.

Do not refer to the text when taking this test.
In questions 1–4 match the following terms:

    a. thymus gland   b. spleen
    c. lymph node   d. red bone marrow

____ 1. causes maturation of T cells

____ 2. purifies lymph

____ 3. purifies blood

____ 4. formation of various types of white blood cells

____ 5. Lymph nodes house
    a. neutrophils and monocytes.
    b. lymphocytes and macrophages.
    c. granular leukocytes.
    d. red blood cells.

____ 6. The spleen
    a. contains stem cells from the bone marrow.
    b. is located along the trachea.
    c. produces a hormone believed to stimulate the immune system.
    d. contains red pulp and white pulp.

____ 7. The thymus
    a. contains all types of stem cells from the bone marrow.
    b. is located along the trachea.
    c. produces a hormone believed to stimulate the immune system.
    d. Both *b* and *c* are correct.

____ 8. All of the following are functions of macrophages EXCEPT
    a. release a colony-stimulating factors to increase the production of white blood cells.
    b. phagocytizing bacteria.
    c. scavenging dead and decaying tissue.
    d. transporting oxygen in the blood.

____ 9. Activity of the complement system is an example of nonspecific defense by
    a. barriers to entry.
    b. phagocytic cells.
    c. protective proteins.
    d. Both *a* and *c* are correct.

____ 10. Secretions of the oil glands is an example of non-specific defense by a(n)
    a. barrier to entry.
    b. protective protein.
    c. phagocytic cell.
    d. acidic pH.

____ 11. Interferon is produced by cells in response to the presence of
    a. chemical irritants.
    b. viruses.
    c. a bacterial infection.
    d. a malarial parasite in blood.

____ 12. The most active white blood cell phagocytes are
    a. neutrophils and monocytes.
    b. neutrophils and eosinophils.
    c. lymphocytes and monocytes.
    d. lymphocytes and neutrophils.

____ 13. The white blood cells primarily responsible for specific immunity are
    a. neutrophils.
    b. eosinophils.
    c. macrophages.
    d. lymphocytes.

____ 14. Which of these is NOT a valid contrast between T cells and B cells?

T cells	B cells

    a. matures in the thymus—matures in bone marrow
    b. antibody-mediated immunity—cell-mediated immunity
    c. antigen must be presented by APC—direct recognition
    d. cytokines—do not produce cytokines

____ 15. A particular antibody can
    a. attack any type of antigen.
    b. attack only a specific type of antigen.
    c. be produced by any B cell.
    d. be produced by any T cell.

____16. The clonal selection theory refers to the
   a. presence of four different types of T cells in the blood.
   b. response of only one type of B cell to a specific antigen.
   c. occurrence of many types of plasma cells, each producing many types of antigens.
____17. The portions of an antibody molecule that pair up with the foreign antigens are the
   a. heavy chains.
   b. light chains.
   c. variable regions.
   d. constant regions.
____18. Which of these is incorrect?
   a. helper T cells—orchestrate the immune response
   b. cytotoxic T cells—stimulate B cells to produce antibodies
   c. memory T cells—long-lasting active immunity
   d. suppressor T cells—shut down the immune response

____19. A person receiving an injection of gamma globulin as a protection against hepatitis is an example of
   a. naturally acquired active immunity.
   b. naturally acquired passive immunity.
   c. artificially acquired passive immunity.
   d. artificially acquired active immunity.
____20. A person vaccinated to produce immunity to the flu is an example of
   a. naturally acquired active immunity.
   b. naturally acquired passive immunity.
   c. artificially acquired passive immunity.
   d. artificially acquired active immunity.
____21. Allergies are caused by
   a. strong toxins in the environment.
   b. autoimmune diseases.
   c. the overproduction of IgE.
   d. the receipt of IgA in breast milk.
____22. Which is NOT true of an autoimmune response?
   a. responsible for such diseases as multiple sclerosis, myasthenia gravis, and perhaps type 1 diabetes
   b. occurs when self-antibodies attack self-tissues
   c. interferes with the transplantation of organs between one person and another
   d. All of these are correct.

THOUGHT QUESTIONS

Use the space provided to answer these questions in complete sentences.

23. How do we know that one aspect of specific immunity is the ability of the body to recognize self as opposed to nonself?

24. Describe how the lymphatic system aids the activities of the integumentary system, and vice versa.

Test Results: _____ number correct ÷ 24 = _____ × 100 = _____%

STUDY QUESTIONS

1. a. S **b.** P **c.** P **d.** S **2. a.** S **b.** RBM **c.** T **d.** S **3. a.** T **b.** RBM **c.** S **d.** L **e.** T **4. a.** in pharynx **b.** intestinal wall **c.** attached to the cecum **5. a.** 2 **b.** 1 **c.** 3 **d.** 1 **e.** 2 **f.** 2 **g.** 1 **h.** 3 **i.** 2 **j.** 1 **k.** 4

6.

Cytotoxic T Cell		B Cell	
a.	yes	b.	yes
c.	yes	d.	no
e.	yes	f.	yes
g.	yes	h.	no
i.	no	j.	yes

7. a **8.** antigens select the B cell that will clone **9.** c **10. a.** remain in body ready to produce more antibodies when needed **b.** undergo apoptosis and die off **11. a.** antigen-binding site **b.** variable region **c.** constant region **d.** heavy chain **e.** light chain **f.** combine with antigens and mark them for destruction **12.** a **13.** identify cell as belonging to a particular individual and make organ transplants difficult. **14. a.** secretes cytokines that stimulate other immune cells **b.** destroys nonself protein-bearing cells, such as virus-infected cells or cancer cells. **c.** undergo apoptosis and die **15. a.–e.** See Fig. 21.9b, p. 424, in text. **f.** immunological memory **g.** memory B **h.** memory T **16.** passive immunity **17.** tetanus, pneumococcal, diphtheria, hepatitis B, measles, mumps, rubella, polio, *Haemophilus influenzae* (type b), chicken pox **18.** It is better to prevent a disease than to try to treat it. **19. a.** Interleukins activate and maintain killer activity of T cells.

b. Interferon causes other cells to resist a viral infection. **20.** Same-type antibodies produced by the same B cell. **21. a.** T **b.** T **c.** T **22. a.** Antigen attaches to IgE antibodies on mast cells and histamine release causes allergic response. **b.** Antibodies and cytotoxic T cells attack foreign antigens. **c.** An infection tricks immune cells into attacking tissues of self.

DEFINITIONS WORDSEARCH

```
      E N I M A T S I H
      G     C O M P L E M E N T
  E A N T I B O D Y       A
  N H A N T I G E N   S   U
  I P     A       O   P   T
  K O     P       R   L   O
  O R     O       E   E   I
  T C     P       F   E   M
  Y A     T       R   N   M
  C M     O       E       U
          S       T       N
          I       N       E
          S       I
```

a. antigen **b.** complement **c.** autoimmune **d.** apoptosis **e.** cytokine **f.** macrophage **g.** interferon **h.** antibody **i.** histamine **j.** spleen

CHAPTER TEST

1. a **2.** c **3.** b **4.** d **5.** b **6.** d **7.** d **8.** d **9.** c **10.** a **11.** b **12.** a **13.** d **14.** b **15.** b **16.** b **17.** c **18.** b **19.** c **20.** d **21.** c **22.** c **23.** Ordinarily, antibodies and T cells attack only foreign antigens. If and when they attack the body's own cells, illness results. **24.** The lymphatic system drains excess tissue fluid from the skin and protects against infections. The skin serves as a barrier to entry of pathogens.

22

PARASITES AND PATHOGENS

While many persons think of microbes only in terms of disease, they actually perform various services for us. Only those microbes that have virulence factors such as toxins, adhesion factors, and invasive mechanisms are pathogens and can cause disease. This chapter studies viruses, bacteria, and some types of fungi, protozoans, and worms that cause disease. The study of diseases is critical today because of many current problems such as emerging diseases, bacterial resistance to antibiotics, and the use of microbes as biological weapons.

Make up a chart that lists the diseases discussed in the chapter. Your chart should also include the name of the causative agent, the symptoms, and any other important points you should remember.

Make sure also that you have a firm understanding of the differences between viruses and bacteria and why antibiotics only work against bacteria. Other important topics in this chapter are emerging diseases (p. 440); structure and action of prions (p. 439); the difference between Gram-positive and Gram-negative bacteria (p. 441); and why some bacteria have become resistant to antibiotics (p. 445). Only a change in human behavior can stop this growing problem.

The life cycle of *Plasmodium vivax* (Fig. 22.16, p. 447), the causative agent for malaria, and *Schistosoma* (p. 449), the causative agent for schistosomiasis, are complex. You should be able to understand the life cycles of each of these organisms.

TIP: Be sure to visit the Online Learning Center that accompanies *Human Biology* 9/e. It has practice quizzes, interactive activities, labeling exercises, art quizzes, animations, flash cards, and much more. http://www.mhhe.com/maderhuman9

STUDY QUESTIONS

Study the text section by section. Answer the study questions so that you can fulfill the learning objectives for each section.

22.1 MICROBES AND YOU (PAGES 432–435)

After you have answered the questions for this section, you should be able to
- Give examples to show that bacteria make ecological contributions, economic contributions, and contributions to our health.
- Discuss three types of virulence factors and why they contribute to disease.
- Distinguish between two types of infectious diseases, four modes of transmission, and four disease categories.

1. Make up a sentence including the terms "decomposers" and "inorganic nutrients" that describes the ecological contribution of microbes. _____

2. Three different types of food that bacteria help us to produce are a._____,
   b._____, and c._____. Two other different types of chemicals that bacteria help us to produce are d._____ and e._____.

3. Bacteria contribute to our health. The normal _____ within an organ protect us from invasion by pathogenic bacteria.

4. Place the appropriate letter standing for a virulence factor next to the following statements.

   A–adhesion factor   *C*–capsule   *F*–flagella   *I*–invasive factor   *S*–survival mechanism   *T*–toxin

   a. _____ Some bacteria adhere to cells and colonize tissues.
   b. _____ Breakdown of the ground substance of connective tissue, allowing microorganisms to invade underlying tissue.
   c. _____ Step on a rusty nail, the *Clostridium tetani* stays at the site, but you can develop "lockjaw."
   d. _____ Protects the bacteria from phagocytosis.
   e. _____ Allows bacteria to migrate through mucous layer of the small intestine.
   f. _____ Ability to multiply in the phagocytic cells that are supposed to destroy them.

5. What is the difference between an acute infection and a chronic infection? _____
   _____

6. Using the following alphabetized list of terms, label the four modes of transmission of a disease.

   airborne
   direct contact
   vector-borne
   vehicle

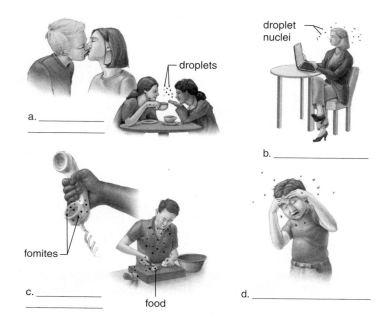

a. _____
b. _____
c. _____
d. _____

droplets
droplet nuclei
fomites
food

7. a. Of a sporadic, pandemic, and epidemic disease, which is most widespread? _____
   b. What is an endemic disease? _____

22.2 VIRUSES AS INFECTIOUS AGENTS (PAGES 436–439)

After you have answered the questions for this section, you should be able to
- Describe and identify the parts of a virus.
- Identify the stages of replication of a virus.
- Discuss influenza as caused by a virus that constantly mutates; two types of herpes as caused by viruses that cause chronic diseases; and measles and the common cold as easily transmitted diseases.
- Describe the basis of prion diseases.

8. In the following diagram, identify three of the four parts of a human virus and fill in the boxes using the following alphabetized list of terms.

      attachment
      biosynthesis
      maturation
      penetration
      release
      replication

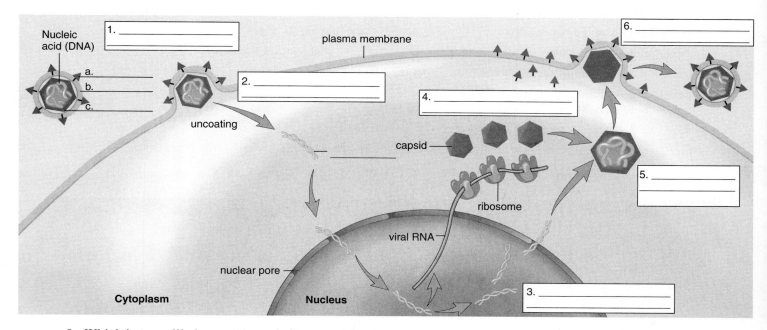

9. Which is more likely to cause an influenza pandemic, an antigenic drift or an antigenic shift? _____

   Why? _____

10. a. Why do herpes viruses cause chronic infections? _____

    b. What is the relationship between chickenpox and shingles? _____

11. Name three ways that you can catch a cold. a._____, b._____, and c._____

12. Indicate whether these statements are true (T) or false (F).
    a. _____ Prions are a special type of virus.
    b. _____ Disease occurs when prions change shape and convert normal prion proteins into the new shape.
    c. _____ Mad cow disease is a prion disease.

22.3 BACTERIA AS INFECTIOUS AGENTS (PAGES 441–444)

After you have answered the questions for this section, you should be able to
• Discuss various types of strep infections and why some are more serious than others.
• Describe a tuberculosis infection and its treatment.
• Distinguish among three common causes of food poisoning.

13. Match the descriptions to the correct strep infection.

    strep throat   scarlet fever   impetigo   necrotizing fasciitis
    a. _____ pharyngitis
    b. _____ cell lysis and muscle tissue destruction on a large scale
    c. _____ relatively mild skin disease
    d. _____ rash that looks like scalded skin

14. *Staphylococcus* lives on the skin or in the nose of _____ people.

15. Tuberculosis can be treated with a._____ , but they must be taken consistently for months or
    years. Lesions in the lungs, called b._____, can persist for years .

16. Tuberculosis is caused by the bacterium a._____ , which produces a(n) b._____
    infection; for example, TB can occur throughout the lifetime of the individual.

17. Label the descriptions with the correct bacteria.
    Salmonella   *Staphylococcus*   *Clostridium botulinum*

    a. _____ long-lasting spores in canned goods

    b. _____ bacteria must grow in body for a time

    c. _____ grows outside the body in foods

22.4 OTHER INFECTIOUS AGENTS (PAGES 446–449)

After you have answered the questions for this section, you should be able to
• Distinguish among three infections caused by fungi.
• Describe the life cycle of *Plasmodium vivax* and the symptoms of malaria.
• Explain why natural surface waters must be boiled before being ingested.
• Contrast the life cycle of and infection caused by the worms *Ascaris* and *Schistosoma*.

18. Cutaneous diseases such as a._____ and b._____, caused by fungi, are
    called c._____.

19. A form of vaginitis and oral thrush are caused by _____ fungi.

20. Over 40 million people have a systemic infection called histoplasmosis, a fungus associated
    with _____ .

21. Indicate whether these statements are true (T) or false (F).
    a. _____ Malaria is the most widespread and dangerous of the diseases caused by protozoans.
    b. _____ A vaccine is now available to cure malaria.
    c. _____ The causative agent of malaria lives in blood cells and causes them to burst, producing chills and
         fever.
    d. _____ Malaria is spread by a vector, the mosquito.

22. Giardiasis is a common a._____ disease caused by a(n) b._____ that lives in water.

23. Place the appropriate letter next to each statement.

    A–Ascaris   *S–Schistosoma*   *N*-neither   *B*–both

    a. _____ Monkey is an intermediary host.
    b. _____ Snail is an intermediary host.
    c. _____ Infection of intestine.
    d. _____ Infection of blood.
    e. _____ Feces contains eggs.

Review key terms by completing this crossword puzzle using the following alphabetized list of terms.

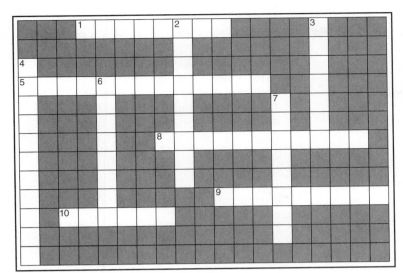

candidiasis
epidemic
Gram stain
helminths
mycosis
pandemic
pathogen
strep throat
toxoplasmosis
vector

Across

1 Disease-causing agent.
5 Disease of animals and humans caused by the parasitic protozoan *Toxoplasma gondii*.
8 An infection caused by *Candida* species commonly involving the skin, mouth, and vagina.
9 Term that designates all parasitic worms.
10 A living organism that transfers disease.

Down

2 A differential staining procedure that divides bacteria into groups based on the cell wall.
3 Any disease caused by a fungus.
4 Disease spread by droplets of saliva or nasal secretions and caused by *Streptococcus*.
6 An increase in the occurrence of a disease within a large and geographically widespread population.
7 Disease that suddenly increases in occurrence above the normal level in a given population.

CHAPTER TEST

OBJECTIVE QUESTIONS

Do not refer to the text when taking this test.

_____ 1. The term *microflora* refers to
   a. pathogenic bacteria within our bodies.
   b. all microbes other than bacteria.
   c. small pathogens.
   d. good bacteria within our bodies.

_____ 2. Microbes that have recently become pathogenic cause _____ diseases.
   a. lethal
   b. emerging
   c. infectious
   d. chronic

_____ 3. Which of the following is NOT a virulence factor?
   a. ability to invade a cell
   b. ability to adhere to surfaces
   c. ability to produce a toxin
   d. All of these are correct.

_____ 4. Which of the following most directly causes disease conditions?
   a. production of a toxin
   b. bacterial growth
   c. bacterial invasion
   d. adherence of bacteria to surfaces

_____ 5. Typhoid fever may result from infection by strains of *Salmonella* that
   a. cause a systemic infection.
   b. move through tissues and into the bloodstream.
   c. are normally found in the colon.
   d. Both *a* and *b* are correct.

_____ 6. An endemic disease
   a. usually affects very few individuals.
   b. is worldwide in occurrence.
   c. is caused by an infectious agent always present in the population.
   d. is controlled through vaccination.

_____ 7. Diseases that generally only last for a short time are
   a. emerging.
   b. endemic.
   c. acute.
   d. sporadic.

_____ 8. A mosquito that transmits malaria is an example of a
   a. vector.
   b. vehicle.
   c. fomite.
   d. pathogen.

_____ 9. The AIDS virus is transmitted by way of
   a. vectors.
   b. indirect contact.
   c. direct contact.
   d. Both *a* and *b* are correct.

_____ 10. During which stage of a viral infection is viral DNA produced?
   a. biosynthesis
   b. maturation
   c. release
   d. replication

_____ 11. Which of the following is NOT a characteristic of viruses?
   a. replicate by way of cell division
   b. have protein capsids
   c. are considered obligate parasites
   d. contain DNA or RNA

_____ 12. What is the function of a viral spike?
   a. replication
   b. attachment
   c. penetration
   d. release

_____ 13. During which step in viral infection are new viruses assembled?
   a. biosynthesis
   b. maturation
   c. release
   d. replication

_____ 14. Which of the following is NOT associated with antigen shift?
   a. large changes in viral character
   b. combination of viral genomes
   c. worldwide infection
   d. yearly outbreaks of influenza

_____ 15. Which virus causes shingles?
   a. herpes simplex type 1
   b. herpes simplex type 2
   c. varicella-zoster
   d. rubeola

_____ 16. Which is NOT a characteristic of prions?
   a. They are small viruses.
   b. They are found in infected nervous tissue.
   c. They are transmitted by ingesting infected tissues.
   d. They cause mad cow disease.

_____ 17. Which term is NOT related to bacterial life?
   a. binary fission
   b. cell wall
   c. nucleus
   d. prokaryotic

_____ 18. Which is NOT a bacterial disease?
   a. common cold
   b. food poisoning
   c. strep infection
   d. tuberculosis

_____ 19. A mycosis is a disease caused by _____ infection.
   a. bacterial
   b. fungal
   c. protozoan
   d. viral

_____ 20. Malaria is caused by a(n) _____ pathogen.
   a. bacterial
   b. fungal
   c. insect
   d. protozoan

____21. How do penicillin antibiotics work?
  a. prevent bacterial protein synthesis
  b. disrupt bacterial cell wall production
  c. prevent bacterial DNA replication
  d. prevent bacterial ribosome activity
____22. Which is caused by a helminth?
  a. toxoplasmosis
  b. schistosomiasis
  c. giardiasis
  d. Both *a* and *b* are correct.
____23. Latent infections are most often caused by
  a. bacteria.
  b. fungi.

  c. prions.
  d. viruses.
____24. Which of the following is transmitted through the air?
  a. measles
  b. common cold
  c. herpes
  d. Both *a* and *b* are correct.
____25. Tuberculosis is a _____ disease.
  a. bacterial
  b. fungal
  c. protozoan
  d. viral

THOUGHT QUESTIONS

Answer in complete sentences.
26. Why are there fewer fungal antibiotics and antiviral medications than bacterial antibiotics?

27. Antibiotic therapy for a bacterial infection can lead to the development of "secondary infections" such as vaginal yeast infections. Why does this occur?

Test Results: _____ number correct ÷ 27 = _____ × 100 = _____ %

ANSWER KEY

STUDY QUESTIONS

1. Decomposers break down dead organic organisms to inorganic nutrients that cycle back to photosynthetic organisms that produce organic food for the rest of the biosphere. **2. a.** alcoholic beverages **b.** bread **c.** milk products or pickled foods **d.** vitamins **e.** antibiotics **3.** microflora **4. a.** A **b.** I **c.** T **d.** C **e.** F **f.** S **5.** Acute comes on suddenly, is intense, and is cured; chronic recurs now and again and has no longlasting cure. **6. a.** direct contact **b.** airborne **c.** vehicle **d.** vector-borne **7. a.** pandemic **b.** always in the population **8.** See Fig. 22.3, p. 436, of text. **9.** An antigenic shift because the viral antigens have undergone a drastic change leaving no one immune. **10. a.** Herpes viruses can lie latent in the body and make their appearance sporadically. **b.** The virus that causes chickenpox can be latent for many years and then later can become active, causing shingles. **11.** airborne droplets, direct contact, vehicle transmission **12. a.** F **b.** T **c.** T **13. a.** strep throat **b.** necrotizing fasciitis **c.** impetigo **d.** scarlet fever **14.** healthy **15. a.** antibiotics **b.** tubercles **16. a.** *Mycobacterium tuberculosis* **b.** chronic **17. a.** *botulinum* **b.** *Salmonella* **c.** *Staphylococcus* **18. a.** ringworm **b.** athlete's foot **c.** tineas **19.** *Candida* **20.** bird droppings **21. a.** T **b.** F **c.** T **d.** T **22. a.** waterborne **b.** protozoan **23. a.** N **b.** S **c.** A **d.** S **e.** B

DEFINITIONS CROSSWORD

Across:
1. pathogen **5.** toxoplasmosis **8.** candidiasis **9.** helminths **10.** vector

Down:
2. Gram stain **3.** mycosis **4.** strep throat **6.** pandemic **7.** epidemic

CHAPTER TEST

1. d **2.** b **3.** d **4.** a **5.** d **6.** c **7.** c **8.** a **9.** c **10.** d **11.** a **12.** b **13.** b **14.** d **15.** c **16.** a **17.** c **18.** a **19.** b **20.** d **21.** b **22.** b **23.** d **24.** d **25.** a **26.** The cells of fungi, like the cells of humans, are eukaryotic, sharing many characteristics related to both cell structure and function. This amount of similarity makes it difficult to produce antibiotics for fungi that will not also harm human cells. Viruses are very different from human cells; they are acellular. **27.** Human microflora consists of many species of bacteria that harmlessly occupy our bodies. They protect us from infection by preventing other harmful microbes from colonizing our tissues. Antibiotics cause the removal of helpful bacteria and leave tissue open for colonization of harmful microbes.

23

SEXUALLY TRANSMITTED DISEASES

STUDY TIPS

No one expects to contract sexually transmitted diseases (STDs), yet every year millions of people do. Many STDs might be considered annoyances, but others can lead to sterility through pelvic inflammatory disease (PID) in women (p. 454), cancer (p. 458), or even death. Everyone should fully understand the preventative measures (p. 463) to avoid contracting or transmitting STDs.

STDs are caused by a variety of pathogens, including bacteria (pp. 454–456), viruses (pp. 457–458), and others, such as pubic lice (p. 459). Among the most interesting are the viruses and retroviruses, because of the manner in which they take over the machinery of the cell

to make more copies of themselves. The AIDS virus is a retrovirus that invades helper T lymphocytes (p. 460). Make a diagram detailing how viruses enter cells and reproduce. Retroviruses go about the same process in a different manner.

Compile a chart listing the STDs discussed in this chapter, the pathogens causing them, symptoms of the disease, dangerous side effects of infection, and whether the disease can be cured. You might also find it interesting to include a column indicating the estimated number of new cases of the disease each year.

TIP: Be sure to visit the Online Learning Center that accompanies *Human Biology* 9/e. It has practice quizzes, interactive activities, labeling exercises, art quizzes, animations, flash cards, and much more. http://www.mhhe.com/maderhuman9

STUDY QUESTIONS

Study the text section by section. Answer the study questions so that you can fulfill the learning objectives for each section.

23.1 BACTERIAL INFECTIONS (PAGES 454–456)

After you have answered the questions for this section, you should be able to
- Name the causative agent for chlamydia, gonorrhea, and syphilis.
- Describe the symptoms of chlamydia in the male and in the female, and compare this disease to a gonorrheal infection.
- Describe the symptoms of a gonorrheal infection in the male and in the female.
- Define PID, and describe how it affects reproduction.
- List the three stages of syphilis.
- Describe how a newborn can contract gonorrhea, chlamydia, or syphilis.

1. *Chlamydia trachomatis* is the most common cause of _____.

2. Name four ways in which chlamydia is like gonorrhea.

   a. _____   b. _____

   c. _____   d. _____

3. What is the danger of pelvic inflammatory disease (PID)? _____

   _____

4. What organism causes gonorrhea? _____

5. a. What are the symptoms of gonorrhea? _____

   b. Does it spread to other areas of the body? _____

6. What is the cause of syphilis? _____

7. Describe the three stages of syphilis.

   a. primary _____

   b. secondary _____

   c. tertiary _____

8. Which two sexually transmitted diseases caused by bacteria can be passed on to the newborn as it passes

   through the birth canal? a. _____   b. _____

23.2 VIRAL INFECTIONS (PAGES 457–458)

After you have answered the questions for this section, you should be able to
- Name the causative agent for hepatitis, genital herpes, and genital warts.
- Describe the results of a herpes infection and genital warts infection.
- Distinguish between strains of hepatitis, and know which is transmitted through sexual contact.

9. What virus causes genital herpes? a. _____ What are the symptoms of genital

   herpes? b. _____

   Where does the virus reside between outbreaks? c. _____ What problems does this disease

   cause in newborns when they contract it from their mothers during the birth process? d. _____

10. a. What virus causes genital warts? _____

    b. What other human disease is caused by this virus? _____

11. Which form of hepatitis is the one spread mostly by sexual contact? a. _____ Is there a vaccine? b. _____

    What are the symptoms of this disease? c. _____

23.3 OTHER INFECTIONS (PAGE 459)

After you have answered the questions for this section, you should be able to
- Tell how *Trichomonas vaginalis* locomotes and explain that it is sexually transmitted.
- Name diseases caused by *Candida albicans*.
- Explain that pubic lice are sexually transmitted.

12. Place *P* for protozoa, *Y* for yeast, *A* for animals, or *O* for other beside each disease.
    a. _____ bacterial vaginitis
    b. _____ trichomoniasis
    c. _____ candidal vaginitis
    d. _____ pubic lice

13. Place *S* for sexually transmitted or *N* for not sexually transmitted beside each infectious disease. Some may be both.
    a. _____ trichomoniasis
    b. _____ pubic lice
    c. _____ candidal vaginitis
    d. _____ bacterial vaginitis

23.4 AIDS (ACQUIRED IMMUNODEFICIENCY SYNDROME) (PAGES 460–466)

After you have answered the questions for this section, you should be able to
- Name the causative agent for AIDS.
- Discuss the three stages of AIDS and how to prevent its occurrence.
- Describe the structure and life cycle of the HIV virus.
- Discuss the treatment choices of AIDS patients.

14. a. What name is given to the retrovirus that causes AIDS? _____

    b. What specific type of T lymphocyte is the host for the AIDS retrovirus? _____

15. How is the use of intravenous drugs related to acquiring HIV? _____

_____

16. Describe the symptoms of the following stages of an HIV infection:

a. Category A _____

_____

b. Category B _____

_____

c. Category C _____

_____

17. Highly active antiretroviral therapy (HAART) uses the drug AZT, which inhibits ᵃ·_____, and one drug that inhibits protease needed for ᵇ· _____. The experimental vaccines use only a portion of the virus to stimulate the production of ᶜ·_____ by the body.

18. In the following diagram, fill in the boxes describing the reproductive cycle of the HIV virus.

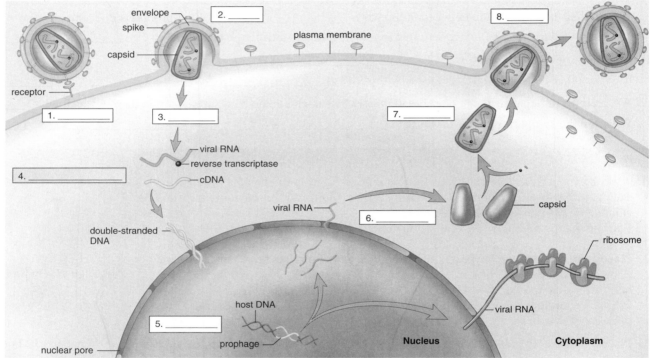

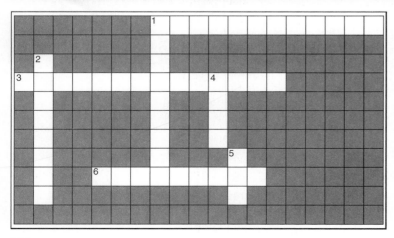

DEFINITIONS CROSSWORD

Review key terms by completing this crossword puzzle using the following alphabetized list of terms.

AIDS
chlamydia
genital warts
gonorrhea
HIV
provirus
trichomoniasis

Across

1 Sexually transmitted disease caused by human papillomavirus, resulting in raised growths on the external genitals.
3 Sexually transmitted disease caused by the parasitic protozoan *Trichomonas vaginalis*.
6 Sexually transmitted disease caused by the bacterium *Chlamydia trachomatis* that can lead to pelvic inflammatory disease.

Down

1 Sexually transmitted disease caused by the bacterium *Neisseria gonorrhoeae* that can lead to pelvic inflammatory disease.
2 HIV DNA that has been integrated into host cell DNA.
4 Disease caused by HIV and transmitted via body fluids.
5 Virus responsible for AIDS.

CHAPTER TEST

OBJECTIVE TEST

Do not refer to the text when taking this test.

____ 1. A retrovirus is
   a. the cause of herpes.
   b. a DNA virus.
   c. capable of reverse transcription.
   d. Both *b* and *c* are correct.

____ 2. Which of these is characteristic of full-blown AIDS?
   a. pneumonia
   b. Kaposi's sarcoma
   c. psychological disturbances
   d. All of these are correct.

____ 3. The cause of AIDS is a
   a. retrovirus.
   b. virus.
   c. bacterium.
   d. protozoan.

____ 4. Which virus causes genital warts?
   a. herpes simplex type 2
   b. human papillomavirus
   c. tobacco mosaic virus
   d. hepatitis B virus

____ 5. Which form of hepatitis is the one contracted sexually?
   a. hepatitis A
   b. hepatitis B
   c. hepatitis C
   d. None are contracted sexually.

____ 6. Which virus causes genital herpes?
   a. herpes simplex type 2
   b. human papillomavirus
   c. tobacco mosaic virus
   d. hepatitis B virus

____ 7. When an infant comes in contact with herpes lesions in his/her mother's birth canal, what complications can be expected?
   a. possible brain damage
   b. possible blindness
   c. possible death
   d. All of these are correct.

____ 8. The name of the organism that causes syphilis is
   a. *Treponema pallidum.*
   b. HIV.
   c. *Neisseria gonorrhoeae.*
   d. *Trichomonas vaginalis.*

____ 9. Congenital gonorrhea is caused by bacteria
   a. crossing the placenta.
   b. in the vagina.
   c. in the oviducts.
   d. All of these are correct.

____ 10. Which phrase best describes most bacteria?
   a. obligate parasite
   b. reproduction inside host cell
   c. cellular organism
   d. Both *a* and *c* are correct.

____ 11. You can cure gonorrhea by
   a. refraining from sexual intercourse.
   b. taking warm baths.
   c. following antibiotic therapy.
   d. taking AZT.

____ 12. Chlamydia is similar to
   a. syphilis.
   b. gonorrhea.
   c. genital herpes.
   d. genital warts.

____ 13. In which stage of syphilis does the victim break out in a rash?
   a. first
   b. second
   c. third
   d. fourth

____ 14. The main complication from prolonged syphilis is
   a. PID.
   b. reduced immunity.
   c. heart and brain involvement.
   d. genital lesions.

____ 15. Syphilis has _____ stage(s).
   a. one
   b. two
   c. three
   d. four

____ 16. Caesarean section is helpful to prevent infection of the newborn with
   a. AIDS.
   b. syphilis.
   c. herpes.
   d. All of these are correct.

____17. Newborns get AIDS
   a. when the virus crosses the placenta.
   b. at birth.
   c. from the mother's milk.
   d. All of these is correct.
____18. Which of these does NOT cause symptoms of vaginitis?
   a. *Chlamydia*
   b. gonorrhea
   c. syphilis
   d. *Trichomonas*
____19. *Candida albicans*
   a. causes vaginitis.
   b. is a normal organism occurring in the vagina.

   c. is a yeast.
   d. All of these are correct.
____20. Pubic lice is a(n)
   a. protozoan.
   b. virus.
   c. animal.
   d. fungus.
____21. Human beings develop active immunity that prevents them from getting sexually transmitted diseases caused by
   a. bacteria.
   b. viruses.
   c. fungi.
   d. None of these is correct.

THOUGHT QUESTIONS

Use the space provided to answer these questions in complete sentences.

22. How does the structure of a virus differ from that of a human cell?

23. What is the benefit of HAART, since there is no cure for AIDS?

Test Results: _____ number correct ÷ 23 = _____ × 100 = _____%

ANSWER KEY

STUDY QUESTIONS

1. sexually transmitted disease   2. a. urethritis b. PID c. cured by antibiotic d. affects newborns   3. it causes sterility in women   4. *Neisseria gonorrhoeae*   5. a. most women are asymptomatic until PID develops; men have painful urination and discharge b. can spread to internal parts, causing heart damage and arthritis; also can cause eye infections   6. *Treponema pallidum*   7. a. chancre b. rash c. gummas   8. a. gonorrhea b. chlamydia   9. a. herpes simplex type 2 virus b. blisters that ulcerate, fever, pain upon urination, swollen lymph nodes c. in ganglia of sensory nerve cells d. grave illness, blindness, neurological disorders, brain damage, and death   10. a. human papillomavirus b. cervical cancer   11. a. hepatitis B b. yes c. flu-like symptoms in 50% of those who contract HBV, can be acute or chronic   12. a. O b. P c. Y d. A   13. a. S b. S, N c. N d. S, N   14. a. HIV (human immunodeficiency virus) b. helper T lymphocytes (cells)   15. Intravenous users often share needles, which directly transfers HIV from one person to the next   16. a. antibodies in blood, swollen lymph nodes b. weight loss, night sweats, fatigue, fever, and diarrhea c. weakness, weight loss, diarrhea, and opportunistic diseases such as Kaposi's sarcoma   17. a. reverse transcriptase b. viral assembly c. antibodies   18. See Fig. 23.12, p. 465, in text.

DEFINITIONS CROSSWORD

Across:
1. genital warts   3. trichomoniasis   6. chlamydia

Down:
1. gonorrhea   2. provirus   4. AIDS   5. HIV

CHAPTER TEST

1. c   2. d   3. a   4. b   5. b   6. a   7. d   8. a   9. b   10. c   11. c   12. b   13. b   14. c   15. c   16. c   17. d   18. c   19. d   20. c   21. d   22. A virus is made up of a capsid and a nucleic acid core, while a human cell has a nucleus and many organelles in a cytoplasm surrounded by a membrane.   23. HAART stops replication of the virus so that the viral load is undetectable.

24

CANCER

STUDY TIPS

You should study this chapter, section by section. The first section considers the characteristics of normal cells versus cancer cells. In normal cells, proto-oncogenes and tumor-suppressor genes function normally to regulate the cell cycle. Also, apoptosis occurs—cells die when DNA is damaged and cannot be repaired. All the characteristics of normal cells are summarized in the Table 24.1 on page 472 in the text.

In cancer cells, proto-oncogenes and tumor-suppressor genes do not function normally due to mutations and the cell cycle is not regulated as before. (Mutated proto-oncogenes are called oncogenes.) Also, apoptosis, normally controlled by a tumor-suppressor gene (for example, *p53*), does not occur. Therefore, a tumor develops. All the characteristics of cancer cells are summarized in Table 24.1 on page 472.

In the second section, which begins on page 476, you learn that heredity and environmental factors lead to cancer. Three types of environmental factors are discussed on pages 476–477: radiation, organic chemicals, and viruses.

The third section, which begins on page 479, is divided into two portions: diagnosis and treatment. Pages 479–481 tell you about the ways to detect the presence of cancer. Everyone should do the home examination tests for breast cancer or testicular cancer, and most people know about the benefits of Pap smears, mammograms, and stool blood tests for colon cancer. More sophisticated are tumor marker tests and genetic tests for oncogenes and mutated tumor-suppressor genes, which have to be done in a laboratory. Various imaging techniques help confirm the presence of cancer. Pages 481–483 are about the treatment of cancer. Both traditional therapies and future therapies are based on the characteristics of cancer cells studied in the second section. Surgery, radiation, and chemotherapy are the traditional therapies. The future therapies include a cancer vaccine to awaken the immune system, *p53* gene therapy to trigger the death of cancer cells, and the use of drugs to inhibit angiogenesis and metastasis. Finally, use of complementary therapies call upon the body to heal itself.

TIP: Be sure to visit the Online Learning Center that accompanies *Human Biology* 9/e. It has practice quizzes, interactive activities, labeling exercises, art quizzes, animations, flash cards, and much more. http://www.mhhe.com/maderhuman9

STUDY QUESTIONS

Study the text section by section. Answer the study questions so that you can fulfill the learning objectives for each section.

24.1 CANCER CELLS (PAGES 472–475)

After you have answered the questions for this section, you should be able to
- Describe the characteristics of normal cells that allow them to participate in the structure and function of a particular body tissue.
- Describe how cell division is regulated in a normal cell.
- Describe why normal cells have limited replicative potential.
- Describe the characteristics of cancer cells that prevent them from participating in the structure and function of a particular body tissue.
- Explain why cell division is unregulated in cancer cells.
- Describe why cancer cells have unlimited replicative potential.

1. Place an X next to the phrases that describe normal cells.
   a. _____ are differentiated
   b. _____ are not differentiated
   c. _____ adhere to their neighbor, form an organized tissue
   d. _____ do not adhere to their neighbor, pile up into an unorganized mass
   e. _____ respond appropriately to growth factors, either stimulatory ones or inhibitory ones
   f. _____ do not respond appropriately to growth factors and keep dividing regardless

2. Place an X next to the phrases that pertain to normal cells.
   a. _____ undergo apoptosis when DNA is damaged
   b. _____ do not undergo apoptosis when DNA is damaged
   c. _____ have an active *p53* tumor-suppressor gene
   d. _____ have an inactive *p53* tumor-suppressor gene
   e. _____ telomeres shorten with each division
   f. _____ telomeres never shorten

3. Place an X next to the phrases that describe cancer cells.
   a. _____ are differentiated
   b. _____ are not differentiated
   c. _____ adhere to their neighbor, form an organized tissue
   d. _____ do not adhere to their neighbor, pile up into an unorganized mass
   e. _____ respond appropriately to growth factors, either stimulatory ones or inhibitory ones
   f. _____ do not respond appropriately to growth factors and keep dividing regardless

4. Place an X next to the phrases that pertain to cancer cells.
   a. _____ undergo apoptosis when DNA is damaged
   b. _____ do not undergo apoptosis when DNA is damaged
   c. _____ have an active *p53* tumor-suppressor gene
   d. _____ have an inactive *p53* tumor-suppressor gene
   e. _____ telomeres shorten with each division
   f. _____ telomeres never shorten

5. Match the frames in the diagram to these descriptions.
   (1) Tumor cells invade lymphatic and blood vessels.
   (2) A cell with a second mutation forms a tumor.
   (3) Cancer in situ.
   (4) One cell in a tissue mutates.
   (5) A new tumor forms at a distant location.
   (6) The tumor invades underlying tissues.

a. _____
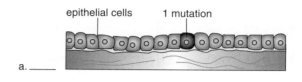

b. _____
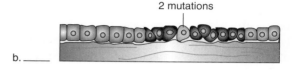

c. _____
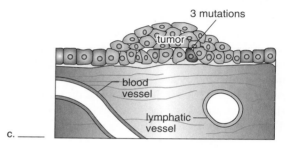

d. _____
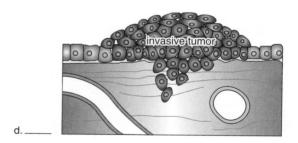

e. _____
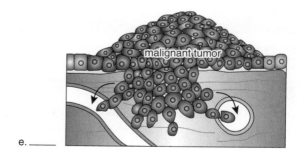

f. _____
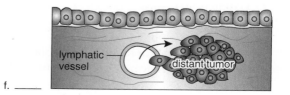

6. In general, state the function of these two types of genes with regard to regulating cell division in normal cells.
   a. proto-oncogenes _____
   _____
   b. tumor-suppressor genes _____
   _____

7. In general, state how these two types of genes affect cell division in cancer cells.
   a. oncogenes _____
   _____
   b. mutated tumor-suppressor genes _____
   _____

24.2 CAUSES OF CANCER (PAGES 476–477)

After you have answered the questions for this section, you should be able to
- Discuss two factors that contribute to the development of cancer.
- List several types of carcinogens.

8. What are two factors that contribute to the development of cancer? a. _____
   b. _____

9. Give two examples of types of cancer that have a hereditary basis. _____
   _____

10. a. What is a carcinogen?_____
    b. Name three types of carcinogens._____
    _____
    c. What habit exposes many people to organic chemicals?_____
    d. What medical procedure exposes many people to radiation?_____
    e. Name a virus known to cause cancer of the cervix._____

11. Name three behaviors associated with the development of cancer. _____
    _____

24.3 DIAGNOSIS AND TREATMENT (PAGES 479–483)

After you have answered the questions for this section, you should be able to
- List and explain routine screening tests.
- Describe tumor marker tests and genetic tests for cancer.
- List and describe ways to confirm the diagnosis of cancer.
- List and describe the standard methods of cancer therapy.
- List and describe future methods of cancer treatment.

12. You would expect these screening tests to detect which type of cancer?
    a. Pap test _____
    b. mammography _____
    c. sigmoidoscopy _____
    d. stool blood test _____

13. Tumor marker tests are a._____ tests that depend upon the fact that the tumor b._____ provoke a(n) c._____ response in the body.

14. Physicians are starting to use genetic tests to help them diagnose cancer. A physician tests for and finds a *ras* oncogene in the urine. In what organ does he/she suspect cancer? a._____ A genetic counselor orders a test for the *BRCA1* gene, and the results come back positive. What type of cancer is the patient apt to have in the future? b._____ Another physician relies on the presence of the enzyme telomerase to reason that cancer cells are present. Why? c._____

15. Which of these imaging techniques is useful for analyzing tumors in tissues surrounded by bone? _____
    a. CAT scan b. MRI c. ultrasound

16. Three standard methods of treatment are a._____, b._____, and
    c. _____.

    d. Explain the rationale behind the use of radiation. _____
    _____

    e. Explain the rationale behind the use of chemotherapy. _____
    _____

    f. Explain the rationale behind bone marrow transplants. _____
    _____

17. Identify these treatments as *radiation, chemotherapy,* or *bone marrow transplant.*
    a. _____ Patient's stem cells are harvested and stored before treatment begins.
    b. _____ Proton beams are aimed at the tumor like a rifle bullet hitting the bull's eye of a target.
    c. _____ Taxol is given to patients with ovarian cancers.

18. Identify these statements as immunotherapy (*IT*) or gene therapy (*GT*).
    a. _____ Use of the genetically engineered adenovirus kills cancer cells when it is injected into the body.
    b. _____ Monoclonal antibodies are designed to zero in on the receptor proteins of cancer cells.
    c. _____ Genetically engineered immune cells are reintroduced into the body.
    d. _____ Retrovirus carrying a normal *p53* gene is injected into tumor cells.

DEFINITIONS WORDSEARCH

Review key terms by completing this wordsearch using the following alphabetized list of terms:

```
X G P A P S M A H I F T E N
Q W O R L C A M R M O Y R T
V C A N X L F P A E Z T E D
A D I N O S A U R T U L M G
G H U C R F N R E A G O O H
E T H A D C B O Y S R T L R
R S U R F I F M U T A G E N
C A N C E R A U N A F A T P
D R A I N O R T A S R C N W
S U T N O L Y H A I G K T O
Z A P O P T O S I S Q R A Y
H A M G A N D F A T D O G Z
H C H E M O N C O G E N E L
H T O N A B Y F V I K L U V
```

apoptosis
cancer
carcinogen
metastasis
mutagen
oncogene
telomere
tumor

a. _____ Environmental agent that can produce mutations.

b. _____ Programmed cell death involving a cascade of specific cellular events leading to death and destruction of the cell.

c. _____ Growth of cells derived from one mutated cell.

d. _____ Malignant tumor that metastasizes.

e. _____ Environmental agent that contributes to the start of cancer.

f. _____ Spread of cancer.

g. _____ Cancer-causing gene.

h. _____ Tip of the end of a chromosome.

CHAPTER TEST

OBJECTIVE TEST

Do not refer to the text when taking this test.

____ 1. Which of these is NOT a known carcinogen?
   a. nitrates
   b. radiation
   c. water
   d. tobacco smoke

____ 2. Choose the one true statement.
   a. Only one-hundredth of cancer deaths can be attributed to smoking.
   b. When smoking is combined with alcohol, cancer risk increases.
   c. Eating smoked meats is good for your health.
   d. Sunbathing produces healthy, tanned skin.

____ 3. Which of these is mismatched?
   a. sunlight–skin cancer
   b. high-fat diet–breast cancer
   c. sexual activity–breast cancer
   d. smoking–lung cancer

____ 4. Factors that lead to carcinogenesis
   a. include heredity, radiation, and carcinogens.
   b. are mostly unknown.
   c. include unsanitary conditions first and foremost.
   d. are highly correlated with drug use.

____ 5. Which of these might explain why vitamins A and C help prevent cancer?
   a. They kill bacteria in the gut.
   b. They are antioxidants.
   c. They cause the growth of cells.
   d. They kill cells.

____ 6. A healthy, cancer-preventing diet should NOT include much/many
   a. vegetables.
   b. fruits.
   c. fiber.
   d. fats.

____ 7. Cancer inhibitory drug therapy includes
   a. monoclonal antibodies to zero in the plasma membrane receptors of cancer cells.
   b. viruses to carry tumor-suppressor genes into cancer cells.
   c. drugs to prevent metastasis and angiogenesis.
   d. acupuncture, biofeedback, and exotic foods to cure cancer.

____ 8. Which of these is NOT a characteristic of cancer cells?
   a. specialization of structure and function
   b. metastasis
   c. uncontrolled growth
   d. nondifferentiation

____ 9. Metastasis is
   a. involved in the initiation of cancer.
   b. a localized tumor.
   c. cured by the proper diet.
   d. the spread of cancerous cells.

____ 10. The formation of new blood vessels, as in a cancerous growth, is
   a. angiogenesis.
   b. spermatogenesis.
   c. oogenesis.
   d. metastasis.

____ 11. Which of these best describes a cancer vaccine?
   a. broken melanoma cells from different sources stimulate the immune system against melanoma
   b. plasma membrane proteins from cancer cells that are introduced into the body to stimulate the immune system
   c. immune cells that display a tumor's antigens to stimulate cytotoxic T cells
   d. A vaccine can be any one of these.
   e. Both a and c are correct.

12. Oncogenes are
   a. viral genes.
   b. tumor-suppressor genes.
   c. always inherited.
   d. involved in the growth cycle of cells.

13. Proto-oncogenes
   a. can become oncogenes by mutation.
   b. help control the growth of cells.
   c. can be carried by viruses.
   d. All of these are correct.

14. Cancer-causing genes are
   a. often oncogenes.
   b. always involved in the growth of cells.
   c. often mutated tumor-suppressor genes.
   d. All of these are correct.

15. Growth factors
   a. are the cause of cancer antigens.
   b. are always receptors.
   c. are chemical signals.
   d. make cells contract.

16. Which of these is a tumor marker test?
   a. monthly breast exam to find tumors
   b. mammogram for breast cancer
   c. CAT scan to find tumors
   d. blood test for CEA to detect colon cancer

17. If cancer is present,
   a. the stimulatory pathway that stretches from the plasma membrane to the nucleus is active and promoting cell division.
   b. the inhibitory pathway is inactive.
   c. oncogenes and mutated tumor-suppressor genes are probably in the nucleus.
   d. apoptosis is not occurring.
   e. All of these are true.

18. Telomerase is an enzyme that
   a. is active in cancer cells.
   b. restores the telomeres in normal cells.
   c. is active in normal cells.
   d. tears down the telomeres in cancer cells.
   e. Both a and d are true.

19. Which of these is mismatched?
   a. mammography—breast cancer
   b. Pap smear—lung cancer
   c. blood test—prostate cancer
   d. heredity—breast cancer

20. Chemotherapy following surgery is for the purpose of
   a. supplying the body with the chemicals it needs.
   b. catching stray cells that have metastasized or might metastasize.
   c. preventing resistance from occurring.
   d. curing leukemia only.

21. Gene therapy for cancer would
   a. use a virus as a vector.
   b. require bone marrow transplants.
   c. inhibit angiogenesis.
   d. Both a and c are correct.

22. Signaling proteins
   a. are active after a cell receives a growth factor or a growth inhibitory factor.
   b. turn on proto-oncogenes and tumor-suppressor genes.
   c. can be involved in the stimulatory or the inhibitory pathway.
   d. All of these are true.

23. Tumor marker tests are always
   a. reliable.
   b. blood tests.
   c. available for any cancers.
   d. All of these are correct.

24. Which is the rationale for p53 gene therapy?
   a. The gene p53 automatically causes the death of all types of cells, even cancer cells.
   b. The gene p53 causes only the death of cells that have severely damaged DNA, namely cancer cells.
   c. The protein p53 is a transcription factor that turns on proto-oncogenes to turn off the cell cycle.
   d. All of these statements are true.

THOUGHT QUESTIONS

Use the space provided to answer these questions in complete sentences.

25. Explain the manner in which cancer can be an inheritable disorder.

26. Explain the manner in which cancer can be a failure of the immune system.

Test Results: _____ number correct ÷ 26 = _____ × 100 = _____%

ANSWER KEY

STUDY QUESTIONS

1. a, c, e **2.** a, c, e **3.** b, d, f **4.** b, d, f **5. a.** (4) **b.** (2) **c.** (3) **d.** (6) **e.** (1) **f.** (5) **6. a.** When active, proto-oncogenes stimulate the cell cycle. **b.** When active, tumor-suppressor genes inhibit the cell cycle. **7. a.** always promote cell division regardless **b.** fail to prevent cell division regardless **8. a.** mutagens **b.** heredity **9.** some breast cancers, retinoblastoma **10. a.** an environmental agent that can contribute to the development of cancer **b.** certain organic chemicals, radiation, pollutants, dietary fat, viruses **c.** smoking **d.** X rays **e.** papillomavirus **11.** smoking, sunbathing, exposure to radiation and organic chemicals **12. a.** cervical cancer **b.** breast cancer **c.** colon cancer **d.** colon cancer **13. a.** blood **b.** antigens **c.** antibody **14. a.** bladder **b.** breast **c.** Except for fetal cells, only cancer cells contain telomerase. **15.** b. MRI **16. a.** surgery **b.** radiation **c.** chemotherapy **d.** Radiation is mutagenic, and dividing cancer cells are more susceptible to its effects. **e.** Chemotherapy is a way to catch cancer cells that have spread throughout the body. **f.** To allow the use of high doses of chemotherapeutic drugs, bone marrow stem cells are harvested and stored before therapy begins. **17. a.** bone marrow transplant **b.** radiation **c.** chemotherapy **18. a.** GT **b.** IT **c.** IT **d.** GT

DEFINITIONS WORDSEARCH

```
                              E
                   M          R
                   E          E
                   T          M
        C       R  A          O
        A       O  S   M U T A G E N
        R       O  T          T
   C A N C E R  U  A
        I       T  S
        N       I
   A P O P T O S I S
        G
        E    O N C O G E N E
        N
```

a. mutagen **b.** apoptosis **c.** tumor **d.** cancer **e.** carcinogen **f.** metastasis **g.** oncogene **h.** telomere

CHAPTER TEST

1. c **2.** b **3.** c **4.** a **5.** b **6.** d **7.** c **8.** a **9.** d **10.** a **11.** e **12.** d **13.** d **14.** d **15.** c **16.** d **17.** e **18.** a **19.** b **20.** b **21.** a **22.** d **23.** b **24.** b **25.** Cancer is inheritable when parents pass on an oncogene or a mutated tumor-suppressor gene to their children. **26.** Cancer may be a failure of the immune system when T cells fail to recognize cancer cells as non-self cells and therefore do not destroy them.

25

HUMAN EVOLUTION

STUDY TIPS

Describe the origin of life (the first cell) while viewing Figure 25.1. Your description should include references to the primitive atmosphere, energy sources, small organic molecules, macromolecules, protocell, and true cell. Distinguish between the RNA-first hypothesis and the protein-first hypothesis (p. 488).

Be able to state why the sameness of living things is dependent upon common descent and the diversity of life is dependent upon natural selection (pp. 490–492). The evidence for evolution falls into various categories. There is the fossil, biogeographical, anatomical, and biochemical evidence. Choose an interesting example from the text for each of these categories to help you remember how it fits in as evidence for evolution. Refer to the bulleted list on page 492 when studying natural selection and be able to list and discuss these steps in their proper order.

Know the classification categories and be able to recite them from the largest to the smallest and vice versa.

To help in this endeavor, you will find it worthwhile to make up a saying such as "*Daring Karen Pushed Cans Off Friendly Grandmother's Stove.*"

Also pay attention to the bulleted list on page 493, which lists primate characteristics. Refer to Figure 25.5 and be able to explain our evolutionary relationship to all the other types of primates. Be familiar with the term *hominid,* which includes australopithecines and humans. Make a chronological listing of the species of hominids, leading up to modern humans (pp. 496–501). Beside each, list the special traits for the group (e.g., tool user, walked upright, etc.) Then make note of how they were similar to modern humans and how they differed.

Be able to describe the difference between the multiregional continuity hypothesis versus the out-of-Africa hypothesis concerning the evolution of modern humans (p. 499).

TIP: Be sure to visit the Online Learning Center that accompanies *Human Biology* 9/e. It has practice quizzes, interactive activities, labeling exercises, art quizzes, animations, flash cards, and much more. http://www.mhhe.com/maderhuman9

STUDY QUESTIONS

Study the text section by section. Answer the study questions so that you can fulfill the learning objectives for each section.

25.1 ORIGIN OF LIFE (PAGES 488—489)

After you have answered the questions for this section, you should be able to
- Describe the chemical evolution that may have led to the origin of the protocell.
- Distinguish between the RNA-first hypothesis and the protein-first hypothesis.
- Distinguish between the protocell and the true cell.
- State two alternative hypotheses regarding how DNA replication evolved.

1. Use these phrases to label the following drawing:
   small organic molecules   primitive gases   plasma membrane   energy sources   macromolecules   ocean

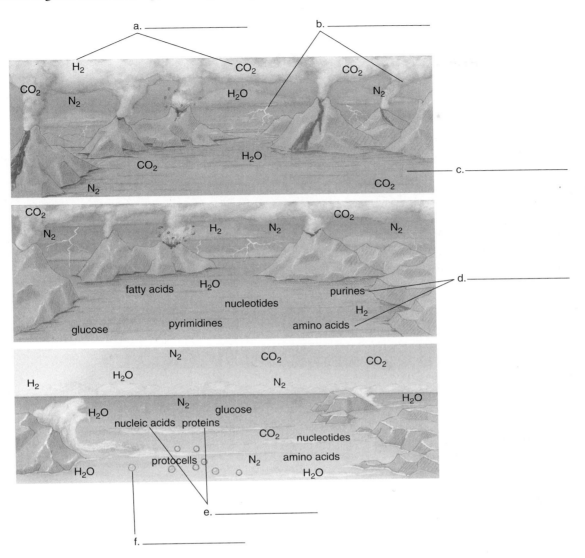

a. _____   b. _____

c. _____

d. _____

e. _____

f. _____

2. The early atmosphere lacked free ᵃ·_____, but there was ᵇ·_____ vapor, which condensed and became the oceans. The gases of the early atmosphere formed ᶜ·_____ when they were exposed to outside ᵈ·_____, such as heat from volcanoes and electric discharges in lightning. The simple organic molecules combined to form ᵉ·_____, and these led to the formation of a protocell, which had a(n) ᶠ·_____. When the protocell could reproduce, it became a true cell, which must have been a(n) ᵍ·_____, living on organic molecules in the ocean and breaking them down anaerobically.

3. DNA replication requires DNA and enzymatic proteins. Fill in this chart to contrast the RNA-first hypothesis with the protein-first hypothesis.

	Cell would already contain	Cell would need to acquire
RNA-first	a. _____	b. _____
Protein-first	c. _____	d. _____

4. a. According to the RNA-first hypothesis, how might the cell acquire DNA and proteins? _____

_____

   b. According to the protein-first hypothesis, how might the cell acquire DNA?

_____

25.2 BIOLOGICAL EVOLUTION (PAGES 489–492)

After you have answered the questions for this section, you should be able to
• Give examples of the evidence for common descent.
• Explain Darwin's mechanism for natural selection, which explains why life is so diverse.
• Explain the sameness and the diversity of living things.

5. Match the following observations to these evidences of evolution:
   history of life   biogeographical   embryological   homologous structures   biochemical
   a. _____ All chordates develop similarly.
   b. _____ Human hemoglobin is like that of a chimpanzee.
   c. _____ A bat wing and a human forearm have the same bones.
   d. _____ Fossils of bacteria are older than fossils of mammals.
   e. _____ Golden jackals in Africa are adapted similarly to coyotes in North America.

6. Number these statements in the proper order to describe the process of natural selection as proposed by Darwin:

   a _____ Each subsequent generation includes more individuals adapted in the same way to the environment.
   b. _____ Resources are limited; therefore, there is a struggle for existence.
   c. _____ Individual members of a species vary in physical characteristics.
   d. _____ Some members of a species are able to capture more resources because they have characteristics that give them an advantage in that environment.

7. Common descent explains the a. _____ of living things and natural selection explains the
   b. _____ of living things.

25.3 HUMANS ARE PRIMATES (PAGES 493–495)

After you have answered the questions for this section, you should be able to
• Name the classification categories in sequence and be able to explain and recognize a binomial name.
• List six characteristics of primates.
• Classify humans.
• Recognize an evolutionary tree.
• Explain our evolutionary relationship to the other primates.

8. a. Make up a saying that will help you remember the order of the classification

   categories._____

   b. What two classification categories are used in a binomial name? _____

_____

9. Place these designations next to the correct category in order to classify humans:

Homo sapiens Mammalia Hominidae Chordata Eukarya Animalia Primates *Homo*

a. _____ domain

b. _____ kingdom

c. _____ phylum

d. _____ class

e. _____ order

f. _____ family

g. _____ genus

h. _____ species

10. Study the following diagram and then fill in the blanks.

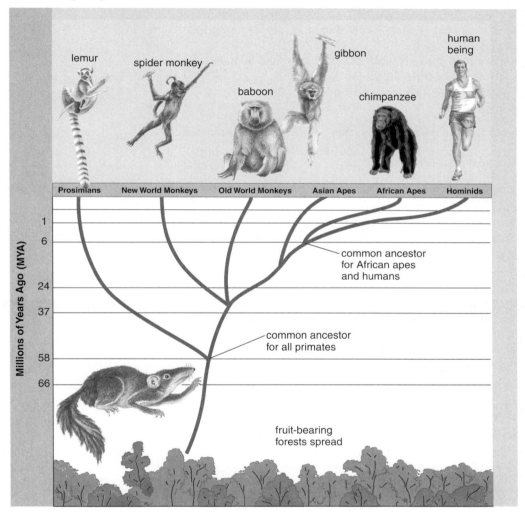

a. All the animals at the top of the illustration are _____.

b. These animals have adaptations for living where? _____

c. Name six characteristics that these animals have in common. _____

_____

d. Which group diverged last from the human line of descent? _____

e. What do you find at a point of divergence? _____

f. What do you call this type of graph? _____

g. What does this type of graph tell you? _____

11. Match the primate characteristics to human features.

opposable thumb   well-developed brain (used twice)   living in trees   long childhood

a. tool use _____ and _____

b. walking erect _____

c. intelligence _____

d. learning after birth _____

25.4 EVOLUTION OF AUSTRALOPITHECINES (PAGE 496)

After you have answered the questions for this section, you should be able to
- Understand and use the term *hominid*.
- Tell where australopithecines evolved.

12. Place a check to indicate which group of animals are all hominids.
    a. _____ all primates
    b. _____ only monkeys, apes, australopithecines, and humans
    c. _____ only apes, australopithecines, and humans
    d. _____ only australopithecines and humans
    e. _____ only humans

Use the space provided to answer this question with a complete sentence.

13. Explain your answer to question 12. _____
    _____

14. Indicate whether these statements about australopithecines are true (T) or false (F).
    a. _____ Australopithecines had a brain as large as modern humans.
    b. _____ Australopithecines were apelike above the waist and humanlike below the waist.
    c. _____ Australopithecines still lived in trees and could not walk erect.
    d. _____ One type of australopithecine is believed to be ancestral to humans.

25.5 EVOLUTION OF HUMANS (PAGES 497–501)

After you have answered the questions for this section, you should be able to
- Distinguish among *Homo habilis, Homo erectus,* Neanderthals, and Cro-Magnon.
- Contrast the multiregional continuity hypothesis with the out-of-Africa hypothesis.
- Give reasons we are all one species despite ethnicity.

15. Arrange these hominids in the order they evolved:

    Homo habilis Neanderthal *Homo erectus* australopithecines Cro-Magnon

    Associate these characteristics with the earliest possible tool use, earliest knowledge of fire, first to walk erect, large brain but primitive appearance, best-developed culture.

    Name   **Characteristic**

    a. _____   _____

    b. _____   _____

    c. _____   _____

    d. _____   _____

    e. _____   _____

16. Place the appropriate letters next to each statement to identify the evolutionary hypothesis.

      MC—multiregional continuity   *OA*—out-of-Africa

    a. _____ *H. erectus* migrated out of Africa into Europe and Asia.

    b. _____ *H. sapiens* migrated out of Africa into Europe and Asia.

    c. _____ *H. sapiens* evolved in Asia, Africa, and Europe.

    d. _____ *H. sapiens* evolved only in Africa.

    e. _____ Humans would have to be more genetically similar.

    f. _____ Humans would be less genetically similar.

In question 17a–b, use the space provided to answer using complete sentences.

17. a. Why do we know now that human ethnic groups represent phenotypes rather than genetically distinct races, as was once thought? _____

_____

    b. To what are these phenotypes adapted? _____

_____

DEFINITIONS WORDSEARCH

Review key terms by completing this wordsearch using the following alphabetized list of terms:

```
A Z F G N P Q L P I R S F
P G N E A N D E R T H A L
R L I T T F S N I E O O R
M V O C U P F P M S M N L
T C R K R B G L A K I V S
A T L E A S T A T G N C T
P O A I L Q T F E Z I V R
S M C L E F B A T R D Q I
B L I S S O F L I F E C Q
B E G G E O C Q E S T L N
C E O F L R I A T B T D S
T I L L E X E G A E N I L
G R O W C A T T M L E I N
T R E A T Y F O I R X R T
G I L F I R A I R K M P P
L L E C O T O R P A S T F
P A T E N T Q I R S B A L
```

fossil
hominid
lineage
natural selection
Neanderthal
primate
protocell
teleological

a. _____ Possible cell forerunner that became a cell once it could reproduce.

b. _____ Process by which organisms become adapted to their environment.

c. _____ Line of descent.

d. _____ Any past evidence of an organism that has been preserved in the Earth's crust.

e. _____ Determined by an end purpose and the outcome is known ahead of time.

f. _____ Culturally advanced hominid with a sturdy build who lived during the last Ice Age in Europe and the Middle East.

g. _____ Member of a family containing humans and australopithecines known only by the fossil record.

h. _____ Member of the order of mammals that includes prosimians, monkeys, apes, and humans.

OBJECTIVE TEST

Do not refer to the text when taking this test.

_____ 1. Which of the following is/are used as evidence of common descent?
   a. fossils
   b. comparative anatomy
   c. comparative biochemistry
   d. All of these are correct.

_____ 2. Which of these statements regarding the fossil record supports evolutionary theory?
   a. Chronologically speaking, fossils exhibit an increase in complexity of form.
   b. Fossils of prehistoric horses reflect a gradual change in form toward the modern horse.
   c. A number of intermediate forms have been found representing transitions between major taxonomic groups.
   d. Specialized features of fossilized organisms correlate well with what environments were believed to be like.
   e. All of these are correct.

_____ 3. In the phrase _theory of evolution,_ the term _theory_ implies
   a. a general, unsupported notion.
   b. absolute fact.
   c. that substantial evidence supports this idea.
   d. None of these is correct.

_____ 4. Related organisms share
   a. similar anatomy.
   b. a common ancestor.
   c. genes.
   d. All of these are correct.

_____ 5. The observation that most species overpopulate their environment best matches with which characteristic?
   a. members of a species vary
   b. struggle for existence
   c. survival of the fittest
   d. adaptation to the environment

_____ 6. Which one of these is NOT a part of Darwin's theory of natural selection?
   a. Organisms strive to develop those traits that are most helpful in their particular environment.
   b. There is severe competition among organisms for available resources.
   c. The members of a species differ from each other.
   d. Better-adapted members of a species will survive and reproduce more successfully.

_____ 7. The observation that whales have flippers best matches with which characteristic?
   a. members of a species vary
   b. struggle for existence
   c. survival of the fittest
   d. adaptation to the environment

_____ 8. The observation that you can tell one domestic cat from another best matches with which characteristic?
   a. members of a species vary
   b. struggle for existence
   c. survival of the fittest
   d. adaptation to the environment

_____ 9. The observation that female lions that are good mothers have the most surviving cubs best matches with which characteristic?
   a. members of a species vary
   b. struggle for existence
   c. survival of the fittest
   d. adaptation to the environment

_____ 10. Which of these is best associated with natural selection?
   a. Both parents pass on their genes to offspring.
   b. Intelligence is a factor of brain size.
   c. The environment determines who will reproduce.
   d. Overpopulation leads to inferior specimens.

_____ 11. When life arose, which was not present?
   a. glucose
   b. water
   c. oxygen
   d. All of these are correct.

_____ 12. The protocell was a(n)
   a. fermenter.
   b. autotroph.
   c. fermenting heterotroph.
   d. fermenting autotroph.

_____ 13. Which pair does NOT give a proper sequence of events?
   a. eukaryotic cell–prokaryotic cell
   b. bacteria evolve–plants evolve
   c. australopithecines–_Homo_ species
   d. primitive gases–protocell

_____ 14. Which of these does NOT describe the primitive atmosphere?
   a. ice cold
   b. no free oxygen
   c. contained gases
   d. no flying creatures

_____ 15. A modern ape, like a human, is a(n)
   a. animal.
   b. mammal.
   c. primate.
   d. All of these are correct.

_____ 16. Which of these are modern humans?
   a. Neanderthals
   b. _Homo erectus_
   c. _Homo habilis_
   d. Cro-Magnon

_____ 17. Which of these are hominids?
  a. great apes
  b. australopithecines
  c. humans
  d. Both *b* and *c* are correct.
_____ 18. *Homo habilis* is a(n)
  a. prosimian.
  b. australopithecine.
  c. human.
  d. All of these are correct.
_____ 19. Which is true of humans but not of apes?
  a. brain size 600 cc or greater
  b. arms shorter than legs
  c. walking upright
  d. All of these are correct.
_____ 20. Which of these is paired with an inappropriate comment?
  a. *Australopithecus africanus*–could be direct ancestor to humans
  b. *Australopithecus robustus*–probably hunted large game
  c. *Australopithecus afarensis*–had a small brain but walked erect
  d. All of these are correct.
_____ 21. Which of these is paired with an inappropriate comment?
  a. *Homo habilis*—tool maker
  b. *Homo erectus*—big-game hunter
  c. *Homo neanderthalis*—remains found only in Africa
  d. Cro-Magnon—drew beautiful pictures
_____ 22. Evidence suggests that
  a. we are descended from one of the modern-day great apes.
  b. we couldn't possibly be related to apes.
  c. we share a common ancestor with African apes.
  d. humans evolved only in Asia and Europe.

THOUGHT QUESTIONS

Use the space provided to answer these questions in complete sentences.

23. Explain why the process of natural selection as proposed by Darwin is nonteleological.

24. How might big-game hunting have influenced the social organization of humans?

Test Results: _____ number correct ÷ 24 = _____ × 100 = _____%

STUDY QUESTIONS

1. a. primitive gases **b.** energy sources **c.** ocean **d.** small organic molecules **e.** macromolecules **f.** plasma membrane   **2. a.** oxygen **b.** water **c.** small organic molecules **d.** energy sources **e.** macromolecules **f.** plasma membrane **g.** heterotroph   **3. a.** RNA **b.** DNA and enzymatic proteins **c.** enzymatic proteins **d.** DNA   **4. a.** RNA genes would code for proteins; reverse transcriptase would make DNA. **b.** Enzymatic proteins are already present and would synthesize DNA from the nucleotides present in the ocean.   **5. a.** embryological **b.** biochemical **c.** homologous structures **d.** history of life **e.** biogeographical **6. a.** 4 **b.** 2 **c.** 1 **d.** 3   **7. a.** unity (sameness) **b.** diversity **8. a.** Example: Daring Karen pushed cans off friendly grandmother's stove. **b.** genus and species   **9. a.** Eukarya **b.** Animalia **c.** Chordata **d.** Mammalia **e.** Primates **f.** Hominidae **g.** *Homo* **h.** *Homo sapiens*   **10. a.** primates **b.** in trees **c.** opposable thumb, well-developed brain, nails, single birth, extended period of parental care, learned behavior. **d.** African apes **e.** common ancestor **f.** evolutionary tree. **g.** the past evolutionary history of a group of organisms, in this case primates   **11. a.** opposable thumb, well-developed brain **b.** living in trees **c.** well-developed brain **d.** long childhood   **12.** d **13.** Hominids are in the family Hominidae and this classification category includes two genera: *Australopithecus* and *Homo*.   **14. a.** F **b.** T **c.** F **d.** T   **15. a.** australopithecines, first to walk erect **b.** *Homo habilis,* earliest possible tool use **c.** *Homo erectus,* earliest knowledge of fire **d.** Neanderthal, large brain but primitive appearance **e.** Cro-Magnon, best-developed culture **16. a.** MC, OA **b.** OA **c.** MC **d.** OA **e.** OA **f.** MC   **17. a.** Molecular data show that there is just as much variation within ethnic groups as there is between ethnic groups. No definite molecular pattern of racial characteristics is evident. **b.** The phenotypes are adapted to climate.

DEFINITIONS WORDSEARCH

```
          N        P
      N E A N D E R T H A L
          T      I   O
          U      M   M
          R      A   I
      L   A      T   N
      A   L      E   I
      C              D
    L I S S O F
      G   E        E
      O   L        T
      L   E    E G A E N I L
      O   C        M
      E   T        I
      L   I        R
    L L E C O T O R P
      T   N
```

a. protocell **b.** natural selection **c.** lineage **d.** fossil **e.** teleological **f.** Neanderthal **g.** hominid **h.** primate

CHAPTER TEST

1. d **2.** e **3.** c **4.** d **5.** b **6.** a **7.** d **8.** a **9.** c **10.** c **11.** c **12.** c **13.** a **14.** a **15.** d **16.** d **17.** d **18.** c **19.** d **20.** b **21.** c **22.** c **23.** Teleological means that the outcome is known ahead of time. The result of natural selection is not predictable because the environment selects the best variation(s) available at the present moment. Variations are dependent on genetic makeup of individuals.   **24.** Big-game hunting seems to have increased social cooperation among humans. Communication between members of the group developed, and the men took the kill home to be shared by all. Big-game hunting may account for division of labor between the sexes. Men went out to hunt, while women stayed home to nurture the young.

26

GLOBAL ECOLOGY

Section 26.1 defines ecological terms, including *ecology, populations, community, ecosystem,* and *biosphere.*

An ecosystem is a community of organisms plus the physical and chemical environment. Some populations are producers (autotrophs) and some are consumers (heterotrophs).

Energy flow and chemical cycling are two important concepts. Use Figure 26.2 to be able to explain energy flow and why ecosystems require a continual supply of solar energy. Chemicals are not lost from an ecosystem; instead, they cycle.

The food webs of ecosystems contain grazing food chains (begin with a producer) and detrital food chains (begin with detritus). Practice constructing food chains for both of these and be able to identify producers and different types of consumers. A trophic level includes all the organisms that feed at a particular link in food chains. Con-

struct an ecological food pyramid using the trophic levels from your food chains. Use the pyramid to illustrate that only 10% of energy is found from one population to the other (Fig. 26.6).

The global cycling of inorganic elements involves the biotic and abiotic parts of an ecosystem. Be able to explain the illustration on the first column of page 510 because it helps you understand the biogeochemical cycles that follow in the chapter. Study each cycle (water, phosphorus, nitrogen, and carbon), and be able to explain each of the arrows in these cycles.

In the water cycle, evaporation of ocean waters and transpiration from plants contributes to aerial moisture. Rainfall over land results in bodies of fresh water plus ground water. Eventually all water returns to the oceans. Be able to discuss global warming, acid deposition, and water pollution and how they relate to the cycles.

TIP: Be sure to visit the Online Learning Center that accompanies *Human Biology* 9/e. It has practice quizzes, interactive activities, labeling exercises, art quizzes, animations, flash cards, and much more. http://www.mhhe.com/maderhuman9

STUDY QUESTIONS

Study the text section by section. Answer the study questions so that you can fulfill the learning objectives for each section.

26.1 THE SCOPE OF ECOLOGY (PAGES 506–508)

After you have answered the questions for this section, you should be able to
- State and define the levels of organization from organism to biosphere.
- Distinguish between producers and consumers and list four different types of consumers.
- Explain the two fundamental characteristics of ecosystems, energy flow and chemical cycling.
- Explain why the same chemicals cycle over and over again in ecosystems.
- Explain why ecosystems require a continual input of solar energy.

1. Match the terms with the following definitions.
   biosphere   community   ecology   ecosystem   population

   a. _____ A community plus its physical habitat.
   b. _____ Members of the same species in the same area.
   c. _____ Study of the interactions of organisms with each other and their surroundings.
   d. _____ The living layer surrounding the Earth.
   e. _____ All the populations interacting in an area.

2. Label this diagram using these terms:

   decomposers   consumers   producers   inorganic nutrient pool

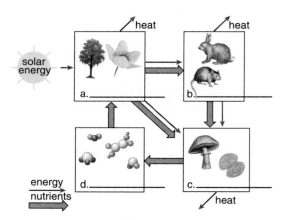

3. Energy doesn't cycle in ecosystems.

   a. Which populations contain the most energy? _____

   b. Which populations contain the least energy? _____

   c. What happens to the energy? Explain on the basis of the second law of thermodynamics. _____

   _____

4. Chemicals do cycle in an ecosystem.

   a. What two molecules do plants use to make glucose? _____

   b. Glucose could conceivably pass from a producer population to the _____

   populations to the _____ populations.

   c. Eventually, the glucose is broken down, and what is returned to plants? _____

   _____

26.2 ENERGY FLOW (PAGES 508–509)

After you have answered the questions for this section, you should be able to
- Describe a food web, and use a food web to construct grazing food chains and detrital food chains.
- Explain an ecological pyramid in terms of trophic levels, and explain why each succeeding feeding level is smaller in biomass/energy than the one before.

This is a food web diagram.

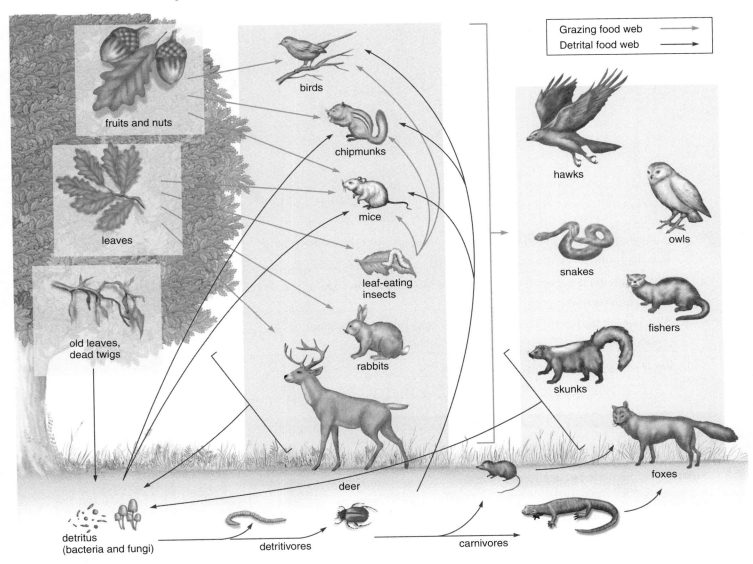

Grazing food web
Detrital food web

fruits and nuts

leaves

old leaves, dead twigs

birds

chipmunks

mice

leaf-eating insects

rabbits

deer

hawks

owls

snakes

fishers

skunks

foxes

detritus (bacteria and fungi)

detritivores

carnivores

5. a. How many trophic levels do you see in the grazing portion of this food web diagram?_____

   b. Name a population at the first trophic level. _____

   c. Name two populations at the second trophic level. _____

   d. Name two populations at the third trophic level. _____

   e. From these trophic levels, construct a grazing food chain. _____

6. Construct a detrital food chain from the food web diagram.

   _____

7. Explain one way in which the detrital food web and the grazing food web are always connected.

   _____

This is an ecological pyramid diagram.

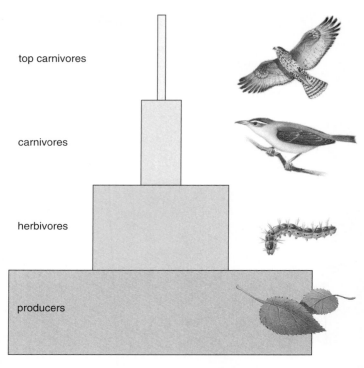

top carnivores

carnivores

herbivores

producers

8. a. In the ecological pyramid, write in two populations at the same trophic level as the caterpillar from the food web diagram. Write the names on this line. _____

    b. In the ecological pyramid, write in two populations at the same trophic level as the bird population. Write these names on this line. _____

9. a. Why is each higher trophic level smaller than the one preceding it? _____

    _____

    b. What is the so-called 10% rule of thumb? _____

    _____

26.3 GLOBAL BIOGEOCHEMICAL CYCLES (PAGES 510–513)

After you have answered the questions for this section, you should be able to
• Construct and give a function for each part of a generalized biogeochemical cycle.
• Describe the water, phosphorus, nitrogen, and carbon biogeochemical cycles.
• Describe the human influence on each of these cycles and the resulting environmental problems that ensue.

10. Examine the following diagram and then answer the questions.

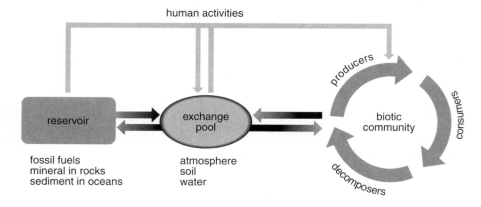

human activities

reservoir

exchange pool

producers

biotic community

consumers

decomposers

fossil fuels
mineral in rocks
sediment in oceans

atmosphere
soil
water

a. What is a reservoir? _____

b. What is an exchange pool? _____

c. What is a biotic community? _____

d. Explain the arrows labeled *human activities*. _____

_____

11. Complete this diagram of the water cycle by filling in the boxes using these terms:

ice H_2O in the atmosphere   ocean   groundwaters

Using these terms, label the arrows:

precipitation (twice)   transpiration from plants and evaporation from soil   evaporation
transport of water vapor by wind

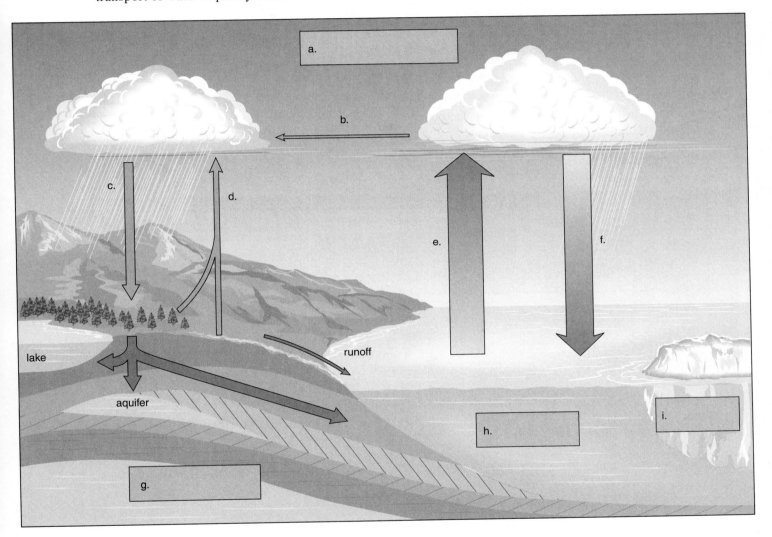

12. Place a check beside the statements that correctly describe the water cycle.
    a. _____ Water cycles between the land, the atmosphere, and the ocean, and vice versa.
    b. _____ We could run out of fresh water.
    c. _____ The ocean receives more precipitation than the land.
    d. _____ Water in the aquifers never reaches the oceans.

13. Place a check beside the statements that correctly describe the results when producers take up phosphate.
    a. _____ becomes a part of phospholipids
    b. _____ becomes a part of ATP
    c. _____ becomes a part of nucleotides
    d. _____ becomes a part of the atmosphere

14. Indicate whether these statements are true (T) or false (F). Rewrite all false statements to be true statements.

    a. _____ Excess phosphate in bodies of water may cause radiation poisoning. Rewrite: _____
    _____

    b. _____ Most ecosystems have plenty of phosphate. Rewrite: _____
    _____

    c. _____ The phosphorus cycle is a sedimentary cycle. Rewrite: _____
    _____

    d. _____ Phosphate enters ecosystems by being taken up by animals. Rewrite: _____
    _____

Questions 15–16 are based on the following diagram of the nitrogen cycle.

15. Match the descriptions to these types of bacteria:
    denitrifying bacteria
    nitrifying bacteria
    nitrogen-fixing bacteria

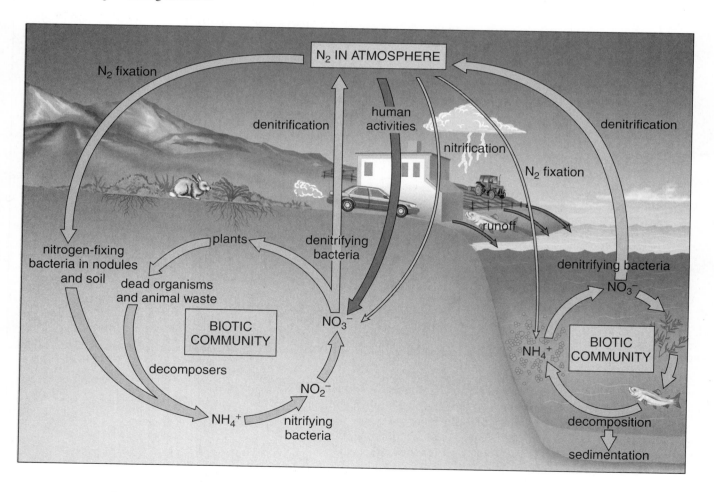

a. _____ bacteria that convert nitrate to nitrogen gas
b. _____ bacteria that convert ammonium to nitrate
c. _____ bacteria in legume nodules that convert nitrogen gas to ammonium

16. Plants cannot use nitrogen gas. What are two ways in which plants receive a supply of nitrogen for incorporation into proteins and nucleic acids? _____

17. When humans produce fertilizers, the gas a._____ is removed from the atmosphere and changed to b._____, which enters the atmosphere.

18. In the carbon cycle, carbon dioxide is removed from the atmosphere by the process of a._____ but is returned to the atmosphere by the process of b._____. Living things and dead matter in soil are carbon c._____ and so are the d._____ because of shell accumulation. In aquatic ecosystems, carbon dioxide from the air combines with water to produce e._____, which algae can use for photosynthesis. In what way do humans alter the transfer rates in the carbon cycle? f._____

26.4 GLOBAL ECOSYSTEMS (PAGES 514–516)

After you have answered the questions for this section, you should be able to
- Describe the two major classes of aquatic ecosystems.
- Explain ecological succession.
- Compare global ecosystems using primary productivity.

19. Aquatic ecosystems can be divided into two major types: the a._____ ecosystems that consist of lakes, ponds, rivers, and streams and the b._____ ecosystems along the coast and in the ocean.

20. Place the following items in the proper order to describe secondary succession: mature forest, annual weeds, shrubs. First, a._____, then b._____, then c._____.

21. For each terrestrial community listed, write a one- or two-word description for the temperature and rainfall.

Ecosystem	Temperature	Rainfall
Tundra		
Desert		
Grassland		
Taiga		
Temperate deciduous forest		
Tropical rain forest		

22. a. What is primary productivity? _____
    b. Which one of the aquatic ecosystems has the highest primary productivity? _____
    c. Which one of the terrestrial ecosystems has the highest primary productivity? _____
    d. Would you expect biodiversity to also be significant in these ecosystems? _____

Review key terms by completing this wordsearch using the following alphabetized list of terms:

```
R G L P T S Q Z I F A G A C N V
B E W D O O F R B A M N C G F D
G U P P Y L R O X A Q U I F E R
F T O D P G M U E F C T D H C L
L R E M U S N O C O R A D R O T
R O S A T D D E F U N N E L S I
P P A N V T N I C H E R P N Y M
N H T E N B T S A X L A O T S J
N I A H C D O O F Q S U S O T E
S C Q B E O R B A U F A I F E R
C A R N I V O R E E O C T I M U
I T F O R D O Q Y P O A I T A M
L I G L U N J U L P H N O L H Z
N O I T A C I F I R T I N E T F
I N N X C B L R Y T S F J R S R
```

carnivore
consumer
ecosystem
food chain
food web
niche
nitrification

a. _____ Role an organism plays in its community, including its habitat and its interactions with other organisms.

b. _____ Order in which one population feeds on another in an ecosystem.

c. _____ Organism that feeds on another organism in a food chain.

d. _____ Consumer in a food chain that eats other animals.

e. _____ Biological community together with the associated abiotic environment; characterized by a flow of energy and a cycling of inorganic nutrients.

f. _____ Process by which nitrogen in ammonia and organic compounds is oxidized to nitrites and nitrates by soil bacteria.

g. _____ In ecosystems, complex pattern of interlocking and crisscrossing food chains.

OBJECTIVE TEST

Do not refer to the text when taking this test.

_____ 1. A complex of interconnected food chains in an ecosystem is called a(an)
  a. ecosystem.
  b. ecological pyramid.
  c. trophic level.
  d. food web.

_____ 2. The biomass of herbivores is smaller than that of producers because
  a. herbivores are not as efficient in converting energy to biomass.
  b. some energy is lost during each energy transformation.
  c. the number of plants is always more than the number of herbivores.
  d. woody plants live longer than omnivores and herbivores.
  e. All of these are correct.

In questions 3–6, indicate whether the statements are true (T) or false (F).

_____ 3. Energy flows through a food chain because it is constantly lost from organic food as heat.

_____ 4. A food web contains many food chains.

_____ 5. The weathering of rocks is one way that phosphate ions are made available to plants.

_____ 6. Respiration returns carbon to the atmosphere.

_____ 7. About _____ of the energy available at a particular trophic level is incorporated into the tissues at the next trophic level.
  a. 1%
  b. 10%
  c. 25%
  d. 50%
  e. 75%

Questions 8–10, refer to the following food chain:
grass → rabbits → snakes → hawks

_____ 8. Each population
  a. is always larger than the one before it.
  b. supports the next level.
  c. is an herbivore.
  d. is a carnivore.

_____ 9. Rabbits are
  a. consumers.
  b. herbivores.
  c. more plentiful than snakes.
  d. All of these are correct.

_____ 10. Hawks
  a. contain phosphate taken up by grass.
  b. give off O_2, which will be taken up by rabbits.
  c. die and decompose, and because of this they cannot contribute to a grazing food chain.
  d. All of these are correct.

_____ 11. Wolves and lions are at the same trophic level because they both
  a. are large mammals.
  b. eat only producers.
  c. eat primary consumers.
  d. live on land.

_____ 12. The form of nitrogen most plants make use of is
  a. atmospheric nitrogen.
  b. nitrogen gas.
  c. organic nitrogen.
  d. nitrates.

_____ 13. Which of the following contribute(s) to the carbon cycle?
  a. respiration
  b. photosynthesis
  c. fossil fuel combustion
  d. All of these are correct.

_____ 14. The largest reserve of unincorporated carbon is in
  a. the soil.
  b. the atmosphere.
  c. the ocean.
  d. lakes.

_____ 15. The greenhouse effect
  a. is caused by particles in the air.
  b. is caused in part by carbon dioxide.
  c. will cause temperatures to increase.
  d. will cause temperatures to decrease.
  e. Both _b_ and _c_ are correct.

THOUGHT QUESTIONS

Use the space provided to answer these questions in complete sentences.

16. Why is a food chain normally limited to four or five links?

17. How would the shortage of an element in the exchange pool affect an ecosystem? Explain.

Test Results: _____ Number right ÷ 17 = _____ × 100 = _____%

STUDY QUESTIONS

1. a. ecosystem **b.** population **c.** ecology **d.** biosphere **e.** community **2. a.** producers **b.** consumers **c.** decomposers **d.** inorganic nutrient pool **3. a.** producers **b.** consumers **c.** It dissipates. With every transformation, as when the energy in food is converted to ATP, there is always a loss of usable energy. Eventually, all solar energy taken in by plants becomes heat. **4. a.** carbon dioxide and water. **b.** consumer, decomposer **c.** carbon dioxide and water **5. a.** three **b.** tree **c.** birds, chipmunks **d.** foxes, fishers **e.** tree → rabbits → snakes **6.** old leaves and dead twigs → bacteria and fungi of decay → earthworms → shrews → hawks **7.** Members of the grazing food web die and are decomposed by bacteria and fungi. **8. a.** rabbits and deer **b.** foxes and snakes **9. a.** Less energy is available to be passed on. **b.** In general, only about 10% of the energy of one trophic level is available to the next trophic level. **10. a.** a source usually unavailable to the biotic community **b.** a source from which organisms do generally take chemicals, such as the atmosphere or soil **c.** producers, consumers, and decomposers that interact through nutrient cycling and energy flow **d.** Humans remove elements from reservoirs and exchange pools and make them available to producers. For example, humans convert N_2 in the air to make fertilizer, and they mine phosphate to make fertilizer. **11.** See Fig. 26.7, p. 510, in text. **12.** a, b, c **13.** a, b, c **14.a.** F . . . may cause algal bloom **b.** F . . . have a limited supply of phosphate **c.** T. **d.** F . . . taken up by plants **15. a.** denitrifying **b.** nitrifying **c.** nitrogen-fixing **16.** nitrogen-fixing bacteria in nodules and nitrate in soil **17. a.** N_2 **b.** N_2O **18. a.** photosynthesis **b.** cellular respiration **c.** reservoirs **d.** oceans **e.** bicarbonate **f.** by burning fossil fuels that add carbon to the atmosphere **19. a.** freshwater **b.** saltwater **20. a.** annual weeds **b.** shrubs **c.** mature forest **21.**

Temperature	Rainfall
cold	little
usually hot	little
moderate	limited
cool	moderate
moderate	moderate
hot	high

22. a. The rate at which producers store energy within organic nutrients. **b.** estuaries, swamps, and marshes **c.** tropical rain forests **d.** yes

DEFINITIONS WORDSEARCH

```
B E W D O O F
                          E
                          C
      R E M U S N O C     O
                          S
            N I C H E     Y
                          S
  N I A H C D O O F       T
                          E
  C A R N I V O R E       M


  N O I T A C I F I R T I N
```

a. niche **b.** food chain **c.** consumer **d.** carnivore **e.** ecosystem **f.** nitrification **g.** food web

CHAPTER TEST

1. d **2.** d **3.** T **4.** T **5.** T **6.** T **7.** b **8.** b **9.** d **10.** a **11.** c **12.** d **13.** d **14.** c **15.** e **16.** By the laws of thermodynamics, energy conversion at each link of a food chain results in nonusable heat. Too little useful energy remains for more links. **17.** A shortage of an element such as nitrogen or phosphorus would reduce the biomass of the producer population. Therefore, the biomass of each succeeding population in the ecosystem would most likely be smaller than it otherwise would be.

27

HUMAN POPULATION, PLANETARY RESOURCES, AND CONSERVATION

Pollution is caused by resource consumption and with more people, there is more resource consumption. There are signs of hope in that the human population is not growing as rapidly as it was previously. Pay attention to and be able to use the boldface terms on page 522 in the text. Study Figure 27.1 and explain its ramifications to a classmate. What are the implications of the age-structure diagram for the LDCs (p. 523)?

Use Figure 27.3 (p. 524) to help you remember the five types of resources discussed in the chapter. Which of these is nonrenewable? Which are renewable? Or does it depend on the specific type?

With regard to your study of land use (pp. 524–525), make sure you can list the drawbacks of human habitation of the beach, semiarid lands, and the tropical rain forest.

The distribution of fresh water on Earth can potentially lead to environmental problems. How do humans increase their ready supply of fresh water, and what are the consequences (pp. 525–526)? What solutions are recommended (p. 526)?

Make sure you can discuss the four characteristics of modern agriculture listed on page 526. What problems are associated with how we usually acquire our food (pp. 526–527)?

It is critical to know the distinction between non-renewable energy sources and renewable energy sources, including the drawbacks of the former and the benefits of the latter (pp. 528–529).

List the causes of extinction in the order of importance and discuss each one in turn (pp. 531–532). Which types of pollution are of most concern to conservationists (p. 531)?

Make a list of indirect values of wildlife and discuss each in turn. Why do conservationists believe that biodiversity is helpful to natural ecosystems (p. 533)? You can more easily observe the various direct values of wildlife. Make a list of the three direct values given on page 535 and give examples of each.

The listing on page 537 offers highlights that should help clarify the chapter for you. Understand the drawbacks of our use of GNP and suggestions for improvement of this indicator of economic health (p. 538).

TIP: Be sure to visit the Online Learning Center that accompanies *Human Biology* 9/e. It has practice quizzes, interactive activities, labeling exercises, art quizzes, animations, flash cards, and much more. http://www.mhhe.com/maderhuman9

Study the text section by section. Answer the study questions so that you can fulfill the learning objectives for each section.

27.1 HUMAN POPULATION GROWTH (PAGES 522–523)

After you have answered the questions for this section, you should be able to
- Describe past, present, and projected population growth.
- Calculate the growth rate of a population given the birthrate and the death rate.
- Distinguish between the more-developed countries and the less-developed countries in terms of population growth and standard of living.
- Use an age-structure diagram to predict future population growth.

1. During the recent period when the human population was undergoing exponential growth, each

year a._____ more/fewer people were added to the population than the previous year.

Therefore, the population growth curve went sharply b._____.

2. At the present time, the growth rate of the population is 1.3%. That means that the birthrate is ^{a.}_____ (higher, lower) than the death rate and that the population will continue to ^{b.}_____.

3. Most of the growth in the human population will take place in the ^{a.}_____ countries, which have a(n) ^{b.}_____-shaped age-structure diagram. This means that ^{c.}_____ young women are entering the reproductive years than older women are leaving them behind.

27.2 HUMAN USE OF RESOURCES AND POLLUTION (PAGES 524–530)

After you have answered the questions for this section, you should be able to
- State the difference between a nonrenewable resource and a renewable resource.
- Give reasons it is unwise to develop beaches, semiarid lands, and tropical rainforests.
- Discuss the availability of fresh water, relative consumption of fresh water, and drawbacks of using dams and taking water from aquifers.
- Describe the benefits and drawbacks of modern farming, animal husbandry, and fishing practices.
- List the nonrenewable sources of energy, and discuss the relationship between fossil fuel consumption and global climate change.
- List the renewable sources of energy, and discuss the expected solar-hydrogen revolution.
- Describe the dangers of mineral consumption, synthetic organic compounds, solid wastes, endocrine-disrupting contaminants, sewage, and agricultural and industrial wastes.

4. Place an X beside the example of renewable resource consumption.
   a. _____ Joe pulled into the gas station and filled up the *gas tank*.
   b. _____ Then he stopped for a *hamburger* on his way home.
   c. _____ Joe had a *shower* before turning in for the night.
   d. _____ He decided to go to sleep while listening to the *radio*.

5. Name two benefits of not developing beaches.
   a. _____
   b. _____

6. When people live on semiarid lands or in deforested tropical rain forests, the ecosystem is often subjected to _____ and becomes useless for human habitation.

In questions 7–9, place an X next to the correct answer.

7. Which of these uses the most fresh water?
   a. _____ industry
   b. _____ agriculture
   c. _____ home use

8. How much of the world's food crop requires irrigation-intensive agriculture?
   a. none
   b. 10%
   c. 40%
   d. 100%

9. In the more-developed countries,
   a. _____ more water is used for bathing, flushing toilets, and watering lawns than for drinking and cooking.
   b. _____ the needs of the human population, in terms of water, are met all over the world.
   c. _____ the world's large dams catch 50% of all precipitation runoff.

10. Name three drawbacks to using dams to get water for irrigation.

    a. _____

    b. _____

    c. _____

11. What are the consequences of removing more water from aquifers than can be recharged by normal means?

    a. _____

    b. _____

12. Name four characteristics of modern farming methods and the potential drawback of each.

    a. _____

    b. _____

    c. _____

    d. _____

13. Name three ecological/societal benefits if all humans ate no more protein than is required to maintain good health.

    a. _____

    b. _____

    c. _____

14. The burning of fossil fuels adds a._____ to the air and is believed to be causing b._____ change.

15. Renewable types of energy include a._____, b._____, c._____, and d._____. Of these, which holds the most promise? e._____

16. What is the solar-hydrogen revolution? _____
    _____

17. a._____ mining of minerals leads to destruction of natural areas, and when metals are disposed of improperly, it leads to human b._____. Synthetic organic compounds found in pesticides, herbicides, and other products are also known to cause c._____.

27.3 BIODIVERSITY (PAGES 531—535)

After you have answered the questions for this section, you should be able to
- Identify the major factors responsible for loss of biodiversity.
- Explain how acid rain, ozone depletion, organic chemicals, and global warming threaten the world's biodiversity.
- Give examples of direct values and indirect values of biodiversity.

18. Match the threats of biodiversity in the alphabetized list to the definition.

    alien species   disease   habitat loss   overexploitation   pollution

    a. _____ The number of individuals taken from a wild population severely reducing its numbers.
    b. _____ Nonnative members of an ecosystem take over and force out the native species.
    c. _____ Wildlife is being exposed to new pathogens.
    d. _____ Human occupation of land destroys the natural ecosystem.
    e. _____ Environmental changes can adversely affect the lives and health of living things.

19. Indicate whether these statements are true (T) or false (F).
    a. _____ A species is often affected by more than one of the threats mentioned in question 18.
    b. _____ Alien species are sometimes brought into new areas by horticulturists.
    c. _____ Tropical rain forest destruction should be associated with the introduction of disease into an area.
    d. _____ Although the interiors of continents have suffered much loss of habitat, coastlines have not suffered as much loss.

20. Match the concerns to the type of pollution:
    acid rain   ozone depletion   organic chemicals   global warming
    a. _____ dying forests
    b. _____ death of coral reefs and rise in sea levels
    c. _____ possible hormonal effects of pesticides and detergents
    d. _____ stunted crop and tree growth

21. Place an X beside all those areas NOT dependent on biodiversity.
    a. _____ ecotourism
    b. _____ prevention of soil erosion
    c. _____ biogeochemical cycles
    d. _____ provision of fresh water
    e. _____ waste disposal
    f. _____ regulation of climate

22. Give an example of the value of biodiversity to:
    a. medicine _____

    b. agriculture _____

    c. consumptive use _____

23. Based on your answer to question 22, the conclusion is that biodiversity has _____ (*much* or *little*) indirect value.

27.4 WORKING TOWARD A SUSTAINABLE SOCIETY (PAGES 537–538)

After you have answered the questions for this section, you should be able to
- List the characteristics that make today's society unsustainable.
- List the characteristics of a sustainable society.
- Tell what the GNP now measures and what it should also measure.

24. Place an X beside those features that characterize today's unsustainable society.
    a. _____ destruction of natural ecosystems
    b. _____ agricultural practices that are wasteful and cause pollution and human illness
    c. _____ grow crops to feed animals
    d. _____ overuse of fresh water from aquifers
    e. _____ dependence on fossil fuel energy
    f. _____ wasteful use of minerals, the mining of which causes environmental degradation and the disposal of which can cause human illness

25. In general, our "throwaway" society is characterized by a high a._____ of nonrenewable energy and raw materials and a large output of b._____ and energy in the form of heat. In contrast, a sustainable society would use c._____ energy sources and would d._____ materials to reduce the amount of e._____.

26. How does our society's use of the GNP hinder the development of a sustainable society that would preserve the environment and human health? _____
_____

DEFINITIONS CROSSWORD

Review key terms by completing this crossword puzzle using the following alphabetized list of terms.

biotic potential
deforestation
desertification
fossil
growth rate
minerals
nonrenewable
renewable
resource
sustainable

Across

1 Removal of trees from a forest in a way that ever reduces the size of the forest.
3 Anything with potential use in creating wealth or giving satisfaction.
5 Naturally occurring inorganic substance containing two or more elements.
6 Maximum growth rate of an organism under ideal conditions.
8 Resources normally replaced or replenished by natural processes.
9 Management of an ecosystem so that it maintains itself while providing services to human beings.

Down

1 Denuding and degrading a once-fertile land, causing long-term changes in soil, climate, and biota of an area.
2 Minerals, fossil fuels, and other materials present in essentially fixed amounts in our environment.
4 The difference between the number of persons born per year and the number of persons who die each year.
7 Any past evidence of an organism that has been preserved in the Earth's crust.

CHAPTER TEST

OBJECTIVE QUESTIONS

Do not refer to the text when taking this test.

____ 1. Which of the following describes biotic potential?
   a. maximum population the environment can support
   b. maximum growth rate under ideal conditions
   c. maximum birthrate for a species
   d. maximum resource usage for a population

____ 2. Which is not a characteristic of less-developed countries?
   a. located mainly in Asia, Africa, and Latin America
   b. poor standard of living
   c. natural resource depletion
   d. slow population growth

____ 3. When a population's age structure is stabilized,
   a. the population remains the same.
   b. replacement reproduction leads to zero population growth.
   c. the population declines.
   d. All of these choices can be correct.

____ 4. Which of the following is a renewable resource?
   a. fossil fuel
   b. minerals
   c. water
   d. land

_____ 5. What proportion of the world's population live within 100 kilometers of a coastline?
  a. 20%
  b. 30%
  c. 40%
  d. 50%

_____ 6. Which does not contribute to beach destruction?
  a. erosion
  b. cooler climate
  c. loss of habitat of marine organisms
  d. pollution

_____ 7. The most common cause of desertification is
  a. overgrazing.
  b. drought.
  c. mining.
  d. construction of dams.

_____ 8. Why is deforestation often followed by desertification?
  a. low rainfall in forest regions
  b. poor soil quality in forest regions
  c. decrease in oxygen production following deforestation
  d. accumulation of carbon dioxide following deforestation

_____ 9. Development of rain forests for human habitation leads to
  a. deforestation.
  b. resource depletion.
  c. loss of biodiversity.
  d. All of these are correct.

_____ 10. Which region has the least amount of fresh water available?
  a. northern Africa
  b. western Europe
  c. North America
  d. South America

_____ 11. How is most fresh water used?
  a. drinking and cooking
  b. flushing of toilets
  c. crop irrigation
  d. industrial purposes

_____ 12. Damming of rivers can lead to
  a. water loss through evaporation.
  b. sediment buildup in reservoirs
  c. increased salinity of water.
  d. All of these are correct.

_____ 13. Saltwater intrusion results from
  a. beach erosion.
  b. aquifer depletion.
  c. damming of rivers.
  d. excessive irrigation.

_____ 14. The settling of soil as it drys out due to aquifer drainage is called
  a. environmental inversion.
  b. salinization.
  c. intrusion.
  d. land subsidence.

_____ 15. How does the United States differ from most other more-developed countries?
  a. poorer standard of living
  b. higher population growth rate
  c. higher death rate
  d. All of these are correct.

_____ 16. Why does monoculture agriculture entail greater risk than planting a variety of crops?
  a. Monocultures use more water.
  b. Monocultures require more fertilizer.
  c. Monocultures are vulnerable to a single type of parasite.
  d. Monocultures require large quantities dangerous pesticides.

_____ 17. Salinization of farm land results from
  a. excessive irrigation.
  b. monoculturing.
  c. herbicide usage.
  d. topsoil depletion.

_____ 18. The result of the green revolution was the
  a. protection of species diversity.
  b. conservation of topsoil.
  c. introduction of high-yield crops to less-developed countries.
  d. increased reliance on polyculture agriculture.

_____ 19. How do modern fishing practices threaten biodiversity?
  a. physical destruction of habitat
  b. removal of community's food supply
  c. chemical poisoning of the water
  d. accidental capture of unwanted species
  e. All of these threaten biodiversity.

_____ 20. What prevents nuclear power from supplying a greater proportion of the world's energy needs?
  a. concern about nuclear power dangers
  b. difficulty in storing radioactive wastes
  c. excessive pollutants released into the air from power plants
  d. Both *a* and *b* are correct.

_____ 21. Which energy source supplies the greatest proportion of the renewable energy supply in the United States?
  a. geothermal energy
  b. hydropower
  c. nuclear energy
  d. wind power

_____ 22. Which of the following is a renewable resource?
  a. fossil fuels
  b. sand
  c. copper
  d. solar energy

_____ 23. Which chemical is most closely related to cultural eutrophication?
  a. chlorofluorocarbon
  b. lead
  c. nitrates
  d. oxygen

_____24. Which factor does not contribute to the unsustainability of modern human society?
a. underuse of available land for human purposes
b. fossil fuel usage for agriculture
c. mining of minerals
d. a large amount of meat in the diet

_____25. Calculation of GNP does not take _____ into account.
a. economic activities
b. environmental value
c. the costs of manufacturing
d. the profits of manufacturing

THOUGHT QUESTIONS

Answer in complete sentences.

26. Human land use can often have long-lasting and widely ranging effects. Describe how deforestation to produce grazing land could lead to desertification later on. How could the construction of a dam on a river contribute to beach erosion hundreds of miles away?

27. Could the roast beef sandwich that you may choose for lunch contribute to the destruction of the environment? Think about and list ways that beef and wheat production can adversely affect the environment.

Test Results: _____ number correct ÷ 27 = _____ × 100 = _____ %

ANSWER KEY

STUDY QUESTIONS

1. a. more **b.** upward **2. a.** higher **b.** grow **3. a.** less-developed **b.** pyramid **c.** more **4.** b, c **5. a.** helps prevent beach erosion **b.** preserves habitats, especially for juvenile aquatic species **6.** desertification **7.** b **8.** c **9.** a **10. a.** loss of water due to evaporation and seepage **b.** salt builds up, making rivers too salty **c.** sediment buildup may make dams useless **11. a.** land subsidence **b.** saltwater intrusion **12. a.** planting few genetic varieties—can lead to devastation by one particular parasite **b.** heavy use of fertilizers, pesticides, and herbicides—contribute to water pollution and are harmful to human health **c.** generous irrigation—overuse of water from aquifers **d.** excessive fuel consumption—air pollution **13. a.** less fossil fuel usage **b.** more food for other humans **c.** less water pollution from animal sewage **14. a.** pollutants **b.** global climate **15. a.** hydropower **b.** wind power **c.** geothermal **d.** solar **e.** solar **16.** The use of solar power to produce hydrogen and the use of hydrogen instead of fossil fuels to power vehicles **17. a.** Surface **b.** illness **c.** illness **18. a.** overexploitation **b.** alien species **c.** disease **d.** habitat loss **e.** pollution **19. a.** T **b.** T **c.** F **d.** F **20. a.** acid rain **b.** global warming **c.** organic chemicals **d.** ozone depletion **21.** None should be checked because all are dependent on biodiversity. **22. a.** Rosy periwinkle is a source of a cancer medicine. **b.** Crops are derived from wild species. **c.** Fish are caught in the wild. **23.** much **24.** a, b, c, d, e, f **25. a.** input **b.** waste materials **c.** renewable **d.** recycle **e.** waste materials **26.** The GNP considers only the monetary value of commerce, but it should also consider the human suffering and the environmental degradation caused by commerce. It should also factor in positive achievements, such as preservation of the environment and scientific and cultural achievements.

DEFINITIONS CROSSWORD

Across:
1. deforestation 3. resource 5. minerals 6. biotic potential 8. renewable 9. sustainable

Down:
1. desertification 2. nonrenewable 4. growth rate 7. fossil

1. b **2.** d **3.** d **4.** c **5.** c **6.** b **7.** a **8.** b **9.** d **10.** a **11.** c **12.** d **13.** b **14.** d **15.** b **16.** c **17.** a **18.** c **19.** e **20.** d **21.** b **22.** d **23.** c **24.** a **25.** b **26.** Forest topsoil is often thin and nutrient-poor because most forest nutrients are held in the trees. As soil quality and plant life decrease, the soil's ability to hold water diminishes. The result is the conversion of a forest into a desert. Beaches naturally lose sediment each year; however, this sediment is replaced by sediment carried by rivers. Dams prevent sediment from reaching the sea, and if sediment loss at a beach is greater than sediment arrival, beach erosion occurs.

27. Beef production: Overgrazing can lead to desertification; animal wastes can enter the water supply; animal feed production typically requires large amounts of fossil fuels that pollute the air and contribute to global warming; feeding grain to animals wastes nutrient energy that could benefit numerous humans; nitrates resulting from animal husbandry may contaminate the water supply. Wheat production: monoculture agriculture places crops at risk from individual parasite species; fertilizers, pesticides, and herbicides can enter the water supply; pesticides and herbicides can serve as endocrine-disrupting contaminants; excessive irrigation can deplete aquifers and contribute to land subsidence and/or saltwater intrusion; modern farming techniques contribute to loss of topsoil.

Notes

Notes

Notes